The Enigma of Loch Ness

The Enigma of Loch Ness

Making Sense of a Mystery

HENRY H. BAUER

University of Illinois Press
Urbana and Chicago

Illini Books edition, 1988

2 3 4 5 C P 5 4 3 2 1

This book is printed on acid-free paper.

Library of Congress Cataloging-in-Publication Data

Bauer, Henry H.
The enigma of Loch Ness.

Bibliography: p.
Includes index.
1. Loch Ness monster. I. Title.
QL89.2.L6B378 1986 001.9'44 85-24554
ISBN 0-252-01284-4 (cloth; alk. paper)
ISBN 0-252-06031-8 (paper; alk. paper)

To the pioneers at Loch Ness
Rupert Gould
Constance Whyte
Tim Dinsdale
David James
Robert Rines

and the champions of the sea serpent
A. C. Oudemans
Rupert Gould
Bernard Heuvelmans

At the time, the discovery . . . was hard—not intrinsically, but because its importance and uniqueness were not well recognized. The discovery was hard also because the data were scattered, confusing, in some respects meagre, in others overabundant. To begin with, it was not clear what was most relevant in all that was known.

Horace Freeland Judson

Contents

Illustrations

Preface

Toward the end of 1933 Loch Ness became famous. Newspapers around the world reported that a monstrously large animal had been seen there by many people: an animal of prehistoric appearance, producing massive disturbances in the water and venturing occasionally onto land. The newspapers also reported the judgments of the scientific experts that no such creature any longer existed. That controversy has persisted up to the present, leaving the Loch Ness monster a celebrated unsolved mystery. This book seeks to make sense of that mystery, by showing that the manifold details of the controversy are inevitable consequences of our approach to knowledge—inevitable in part because we look to science for the answer to such a mystery and because science has no certain answer and yet gives one.

I happen to believe that the Loch Ness monsters (or Nessies) are actual living animals, but this book is not an argument to that effect; rather, my task is to explain why the controversy has persisted so long. I make the cases against the monster and for the monster in the first two chapters in order to introduce the points over which disagreement reigns; in subsequent chapters I show how controversy inevitably flows from the nature of the evidence, from the avocation of monster hunting, from society's attitude toward anomalous beliefs and claims, from the nature of science.

The reality of the Loch Ness monster may be questionable, but the reality of the controversy about it is not. This book is predicated on the belief that useful knowledge will come from an examination of the controversy, whether or not a Loch Ness monster exists. I offer this study as a beginning that will, I hope, stimulate others to make deeper analyses. To that end I have included a bibliography of all the significant writings about the matter that I have been able to locate. I also provide a brief chronology of the Loch Ness story (appendix A); readers unfamiliar with the main events might find it useful to scan those pages before or while reading the book.

Of the many people who helped directly or indirectly as I wrote this book, I mention only a few. Most immediately, Tim Dinsdale, whose book first aroused my interest, whose integrity convinced and captivated me, who kept me au courant, and who read early drafts of parts of this book. What errors or infelicities remain must be laid at my door, not at Tim's; in some places I have gone against his advice. I also have had the benefit of most useful comments from many others, among them Ann Heidbreder Eastman, Arthur M. Eastman, David James, Larry Laudan, Roy Mackal, Adrian Shine, and Marcello Truzzi.

Many helped me learn of and obtain the far-flung literature: notably, Jean Berton, Steuart Campbell, Tim Dinsdale, George Eberhart, Bernard Heuvelmans, J. W. (Dick) MacKintosh, Roy Mackal, Ulrich Magin, R. J. M. Rickard, Marcello Truzzi, Ron Westrum, and Constance Whyte; many librarians, at the University of Kentucky and at Virginia Polytechnic Institute and State University; and marvelous research assistants in Carla Knapp, Jean Hammond, and Helen Bauer. The writing and proofreading of the manuscript were aided immeasurably by the willingness and competence and patience of Becky Cox, who turned untidy scrawls into elegantly formatted typescript.

I am grateful as well to many colleagues for providing me an environment rich in intellectual stimulation and even supportive of such ventures as these: preeminently, John D. Wilson, for opportunity offered and encouragement unstinted. I might not have been able to complete this work were it not for the concern and professional skill of three physicians: William T. Hendricks, of Blacksburg; J. Hayden Hollingsworth, of Roanoke; and Kenneth M. Kent, of the National Institutes of Health (later at Georgetown University). Any attempt to express my gratitude to them would clearly be inadequate.

1

The Monster Is a Myth

Humanity's knowledge of natural history was low in fact and high in fancy for so long that it is a daunting task now to separate the one from the other in legends and the older records. On the one hand we have such instances of fact as valuable drugs extracted from folk medicines and magical herbs, or the founding of archaeology through an inspired analysis of Homeric legends; on the other hand there are such clear instances of fancy as the writings about Atlantis or those about past visitations to Earth by intelligent extraterrestrials.

It remains problematic just how much truth there may be in the oral and written lore about serpents and dragons and water monsters. The Bible indicates that serpents and leviathans had symbolic significance; and stories of dragons, loathly worms, and sea serpents appear to draw heavily from the imaginations of our ancestors. Even in modern times sailors have regarded the sighting of a sea serpent as an omen of disaster.[1] In the British Isles some historical basis for this superstition could perhaps be found in the Viking raids to which the natives were subjected. The Viking ship, with its high prow shaped like a long neck and dragon-like head, was deliberately designed to look like a water monster and to strike fear into the onlooker;[2] the approach of these "sea serpents" in fact boded ill for settlements marked by the Vikings as their prey.

In Scotland folklore traditionally associated large aquatic beasts having magical powers and usually malevolent intentions with many bodies of water, including streams and both small and large lochs. Streams traditionally harbored water-kelpies, whereas lochs were inhabited by water-horses and water-bulls.[3] Several "lochs-of-the-beast," Loch-na-Beistie, are found on old maps, and tales of Nessie-like creatures have come from many lochs well into this century.[4] The survival of myth-in-biology is illustrated by the old Highlander quoted by Constance Whyte: asked whether he had seen the monster in Loch Ness, he replied that he had not, but that he had seen the water-horse more than once.[5] In Ireland also the critical analysis appropriate to a scientific era is not uniformly applied to these matters. As recently as the late 1960s small loughs only a few hundred yards in extent and ten to twenty feet deep were conscientiously dragged with nets to find the large animals supposedly seen there by some of the locals[6] (nothing was found, it need hardly be said). Loch Ness itself was subjected to the rite of exorcism as recently as 1973 to rid it of the Monster.[7]

Perhaps the crux of the problem lies in varying notions of knowledge and of reality. Nowadays we are schooled to believe that truth means scientific truth, which can be demonstrated to others in tangible ways; but that is not the traditional view (which persists in many parts of the British Isles, as elsewhere). Constance Whyte, with insight and without being at all pejorative, has reminded us in *More than a Legend* that folklore, legend, and purely subjective knowledge are regarded as entirely real by many of the older generation in the Highlands. When they say that they have seen a water-horse, they may not mean what we would mean by such a declaration, namely, an animal in the water: it could be a mirage or some other illusion. We are told what has been perceived by these individuals, not what "objective reality" they believe to be behind the perception; they do not necessarily even recognize such a distinction.[8] The resulting confusion was pointed out by a contributor to the Proceedings (1953–54) of the Scottish Anthropological and Folklore Society: the old traditions, he said, had led some people to believe the Loch Ness monster to be an actual animal.[9]

Whyte gives several examples of the mythic nature of the beast: the taboo that still discourages naming or mentioning it, and the warnings to children not to play near the water for fear of the water-kelpie.[10] According to an old prophecy, if the Loch Ness monster is harmed, then half of Inverness will burn and the other half will be flooded.[11] Campbell relates that appearances in Loch Morar of Nessie's cousin a-'Mhorag, or Morag, were an omen of death for a member of the Gillies clan, and later also for the MacDonells and the MacDonalds. Morag was a mermaid, a spirit of the loch, rather than the usual kelpie or water-bull.[12]

When I began work on this book I recalled having read long ago in a semi-autobiographical novel that the Loch Ness monster had been an invention of journalists. I wrote to ask the author of that book about the provenance of his story.[13] Here is his reply:

> Dear Mr. Bauer: Your letter of July 8th has only just reached me.
>
> Let me first tell you that . . . [Lester Smith] writes fiction, although it may often appear otherwise. Truth with trimmings might describe it better. I am afraid I don't remember in which of my books the episode you quote may be found, but it was not necessarily the truth. The truth now follows.
>
> In the early 1930s I, with two young partners, ran a publicity service in London. One of the partners was a native of Lossiemouth, Ramsay Macdonald's home, not far from Loch Ness. On returning from a holiday he brought us a small account. A group of hotels catering to tourists in that area wanted publicity and offered a fee of 50 pounds. We accepted. . . . Around the same time we were offered a more important account by [a realtor] . . . in the Okanagan Valley of British Columbia. . . . We were told and I am inclined to believe that he invented the Ogopogo, a legendary creature inhabiting Lake Okanagan. This was corn in Egypt. The Lossiemouth member of the firm then told us that for centuries a legendary creature was supposed to dwell in Loch Ness. We had never heard of it. At that time our "board room" was the saloon bar of a pub just off Trafalgar Square and over several pints of beer we became midwives of the reborn Loch Ness Monster. All we had to do was to arrange for the Monster to be sighted. This we did and the story snowballed. Thousands went north to see it and see it they did. It was, of course, pure hokum. The unwitting parent was the Ogopogo.
>
> The technique is old and effective and has been used through the ages by organised religions. More recently did we not have soldiers wide-eyed on "seeing" the Angels of Mons!
>
> I hope this helps you.

I could hardly believe that serendipity had thus brought me the explanation of the Loch Ness affair. Could Smith's recollection be trusted, nearly fifty years after the event? Was it not implausible that he could not remember in which of his books he had told the story?

In point of fact I discovered that Smith had written under more than one pseudonym and a total of more than forty books, as well as stories, articles, scripts, and so forth, so it was not implausible that he could not remember just where he had used a by-the-way reference to Loch Ness. Over the course of several years I reread Smith's books as I could find them (most were out of print and had never been published in the United States); several references to Okanagan were tantalizing. Finally I found the following, in an account published in 1950: ". . . a great number of persons have sworn to having seen a creature called the Loch Ness Monster, and you may accept my assurance that the same Loch Ness Monster was born in my presence, during a conversation which took place in a London public house, under the shadow of the monument erected to the great Lord Nelson. . . . The Loch Ness Monster . . . was invented for a fee of 150 pounds by an ingenious publicity man employed by hotel-keepers." So Smith's recollection of the events had been the same in 1950 as in 1980, which makes his story all the more convincing—especially since my query to Smith had been based on a misrecollection of what I had read.

Apparently this publicity scheme of Smith and his colleagues—which doubtless involved the press, be it wittingly or unwittingly[14]—proved to be very effective indeed. Loch Ness became the most popular destination in Britain for motorists: a patrolman for the Automobile Association recalled an occasion when 200 cars were drawn up by the side of the loch, and pandemonium would break out every time he pointed to something. Monster-hunting parties became fashionable, and all the hotels in the area were filled over the Christmas season of 1933. Inverness was flood-lit for the first time. On Boxing Day cars formed virtually a continuous line for the twenty miles from Inverness to Fort Augustus. In February 1934 the Inverness Town Council reduced its expenditure for advertising by one-third since the monster was deemed sufficient publicity.[15]

Smith's revelation helps greatly in elucidating the course of events. The Nessie sighting that spurred the flap of 1933 was reported in the *Inverness Courier* on May 2: an unnamed couple had seen a large creature disporting itself in Loch Ness. Rupert Gould's inquiries revealed that these people were, in point of fact, the lessees of the Drumnadrochit Hotel, near Urquhart Bay on the north side of the loch (the side on which the new road had been built).[16] The report had been given to the *Inverness Courier* by a local correspondent, Alex Campbell, who was also the water bailiff at Fort Augustus.

Campbell himself subsequently saw the monster on a number of occasions: on September 7, 1933, at a distance of about 600 yards, "it seemed to be about 30 feet long, and what I took to be the head was fully 5 feet above the surface. . . . The creature . . . seemed to be watching two drifters passing out of the Canal and into Loch Ness; and . . . it kept turning its head and also its body very quickly. . . . I saw this for fully a minute, then the object vanished as if it had sunk out of sight."[17] Whyte quoted a similar description by "A.C.": distance 500–600 yards, length about 30 feet, head and neck about 5 feet long, quick turning of head, instantaneous sinking as two drifters came down the Canal.[18] A few other details differ (for instance, the date given to Whyte was September 22), but then Whyte's talk with Campbell must have been at least several years after the event.[19] More than twenty-five years later Campbell repeated to Tim Dinsdale the essentials of what he had seen: the length of 30 feet, the long neck, the disappearance as the two drifters came along. But again, minor details differed—the event was ascribed to May or June of 1934, for example.[20] Talking with David Cooke, Campbell placed the sighting in May 1934 and remarked that it was the only time he had seen the head and neck of the monster.[21] In the Walt Disney film *Man, Monsters and Mysteries,* Alex Campbell again recounted this sighting.

Small differences in recollection of a date over such a period of time are surely normal; and the variations may only reflect inaccurate reporting of what Campbell said. At any rate Campbell's integrity is in no way in question here—he was highly respected by all who knew him, and his honesty was apparent to the many people who asked him about the monster: Gould, Whyte, Dinsdale, and others. It might be, however, that we have the sort of problem alluded to earlier: a man honestly relating his subjective perceptions, answering the question as put to him, about his sighting of the monster; but the interviewer mistaking this for Campbell's belief about whatever "objective reality" there may have been behind his perception.

Had Campbell always been asked what he thought it was on the water that caused him to see the monster, he might well have shared—as he did with Rupert Gould—the information in his letter of October 28, 1933, to the Ness Fishery Board. Referring to his sighting of September 7, 1933, from which I quoted earlier, Campbell continued:

> Last Friday I was watching the Loch at the same place and about the same time of day. The weather was almost identical—practically calm and the sun shining through a hazy kind of mist. . . . I discovered that what I took to be the Monster was nothing more than a few cormorants, and what seemed to be the head was a cormorant standing in the water and flapping its wings . . . the bodies of the birds were magnified out of all proportion. . . . This mirage-like effect I have often seen on Loch Ness . . . it gives every object—from, say, a gull or a bottle to an empty barrel—a very grotesque appearance provided that such objects are far enough away.[22]

When speaking with Dom Cyril Dieckhoff, Campbell acknowledged that the beast looked bigger: "You know how a sea-gull can look bigger than a duck floating on a calm day." Dieckhoff's diary also notes that Campbell ". . . later modified his statement to say that sun was in his eyes and a misty haze on the water . . . gave impression that for some reason was anxious to minimise what he had previously said and absolutely refused to allow name to be mentioned to anyone, though previously had expressed willingness to give his evidence to any scientific inquirer, though not to the press."[23] Perhaps Campbell was embarrassed that he had been fooled by a mirage and did not want that to become generally known. So, too, he may have made various modifications in his statements over the years in an attempt to avoid embarrassment without actually lying.

Gould mentioned other reports of such misperceptions as Campbell's and drew attention to published remarks that mirages are frequently seen on Loch Ness; the same points have been made by others.[24] More recently Lehn has suggested that the reports of monsters from a number of lakes in the Northern Hemisphere, at comparable latitudes to Loch Ness, lend support to the idea that optical effects are a contributing cause: the variations of air and water temperatures typically produce temperature inversions in spring and summer, and such inversions, especially when the water is calm, produce distortion and magnification of distant objects whether observed by eye or by photography.[25]

Instances when the observed monster turned out to be a log or a sudden gust of wind also were mentioned by Gould.[26] He himself had seen during his visit to the loch many tar barrels floating on the water, debris from the road building which no doubt had led to an unusually large number of shrubs and trees and pieces of both being in the water during those years. Surely, then, the very large number of sightings of "the Monster" in 1933 and 1934, more frequent than ever before or since, reflects that amount of jetsam in the water at the time. Even the careful and prepared observer can be fooled. Tim Dinsdale, on his initial expedition, was not the first to mistake a tree trunk for the monster until he used binoculars to focus on what he saw, and his first filming of the monster turned out to be footage of a disturbance caused by wind and waves around some rocks.[27] The Oxford and Cambridge expedition rowed toward what it thought was the monster and found a half-submerged fir tree. William Owen was nearly fooled by a swimming stag, and he was completely taken in by a flock of geese. Roy Mackal filmed one monster only to see it take to the air and fly away; and his sighting of another three-humped monster speeding along turned out to be a combination of mirage and flying ducks.[28] Ideal conditions for seeing the monster—"Nessie weather" to monster buffs—are glassy calm, warm, and sunny, just the prerequisites for mirages. Captain Fraser's film of 1934 was taken under conditions of heat haze, as were several of the sequences shot by the Loch Ness Investigation Bureau (LNI).[29] The fact that almost all the monster sightings are very brief—a few minutes or less—is also consistent with mirage effects.[30]

The frequency of sightings has been calculated for the Mountain expedition of 1934 and the LNI expeditions of the 1960s: about one sighting every 350 man-hours of watching,[31] or only one in a month of 12-hour days. Of a total of about 3,000 recorded sightings, Mackal rejected as mistaken or doubtful all but 251, and even the latter include such questionable reports as that of St. Columba in A.D. 565 and hearsay from the nineteenth century. Others have estimated that 80–90 percent of the reported sightings are mistakes, failures to properly identify ordinary objects.[32] Over a period of fifty years only a half-dozen still photographs (other than obvious and outright fakes) have been obtained, and it is not difficult to believe that these resulted from very rare instances of particularly animal-like debris, as discussed at length by Burton in *The Elusive Monster.* Indeed, Dinsdale's unique film could be one of Burton's hypothesized mats of vegetation, one of which was actually seen by Alex Campbell. And the successful photographs obtained underwater by the Academy of Applied Science (AAS), it should be noted, represent but a half-dozen frames out of many hundreds of thousands exposed.[33] Since sightings and photos are so infrequent, an explanation in terms of extremely uncommon phenomena is indicated—mirages, say, or water-logged debris of coincidentally relevant shape.

Much has been made of the "surgeon's" photograph (see fig. 2, in chap. 2), particularly since the photographer was well known and respected. But the circumstances of the taking of that picture have never been made entirely clear, and the surgeon refused to give his own opinion of what the picture showed. Perhaps he felt that a sufficient clue was the date on which he took it, April 1.[34] Quite recently it has been claimed that the surgeon admitted to a close friend that his famous photograph was a hoax; and a recent analysis of the photo reaches a similar conclusion.[35]

All of the salient features of the Loch Ness phenomenon, then, can comfortably be explained by the stimulus of legend and of public relations acting on the gullible and expectant visitor, by human misperceptions made more likely by occasional combinations of mirage-like effects working on such natural objects as birds and tree trunks. Or can they?

Notes

1. Rupert T. Gould, *The Case for the Sea-Serpent,* London: Philip Allan, 1930, p. 210.
2. Elliott Snow, "The Great Sea Serpent," *Nature Magazine,* Oct. 1927, pp. 240–43.
3. Alexander Stewart, *'Twixt Ben Nevis and Glencoe,* Edinburgh: William Patterson, 1885, pp. 39–41.

4. Constance Whyte, *More than a Legend,* rev. 3d imp., London: Hamish Hamilton, 1961, pp. 123–38.

5. Quoted in ibid., p. 38.

6. F. W. Holiday, *The Great Orm of Loch Ness,* New York: Avon, 1970, p. 179; F. W. Holiday, *The Dragon and the Disc,* New York: W. W. Norton, 1973, pp. 48–79.

7. Tim Dinsdale, *Project Water Horse,* London: Routledge and Kegan Paul, 1975, p. 171.

8. James McKinley (1895), quoted in Holiday, *Great Orm,* p. 102.

9. Whyte, *More than a Legend,* 1961, p. 139.

10. Ibid., pp. xvii, 63–64, 135.

11. Michael Baillie, quoted in *Aberdeen Press and Journal,* 12 Oct. 1959.

12. Elizabeth Montgomery Campbell and David Solomon, *The Search for Morag,* London: Tom Stacey, 1972, p. 81.

13. The author, whom I shall call Lester Smith, asked me not to print his well-known nom de plume: "When the book of mine you read appeared I was inundated with letters from all over the world. I don't want that to happen again."

14. In 1959 the London correspondent of an Italian newspaper claimed to have authored the legend of Nessie in the summer of 1933. See "Je suis le père du monstre du Loch Ness," *Match* (Paris), no. 522, 11 Apr. 1959.

15. Nicholas Witchell, *The Loch Ness Story,* Lavenham (Suffolk): Terence Dalton, 1974, pp. 62, 65, 74.

16. Rupert T. Gould, *The Loch Ness Monster and Others,* New York: University Books, 1969, p. 40.

17. Quoted in ibid., pp. 110–11.

18. Whyte, *More than a Legend,* 1961, p. 203.

19. Witchell, *Loch Ness Story,* 1974, p. 117.

20. Tim Dinsdale, *Loch Ness Monster,* London: Routledge and Kegan Paul, 1961, pp. 4, 125.

21. David C. Cooke and Yvonne Cooke, *The Great Monster Hunt,* New York: W. W. Norton, 1969, p. 99.

22. Quoted in Gould, *Loch Ness Monster,* 1969, p. 111.

23. Quoted in Whyte, *More than a Legend,* 1961, pp. 203–4, 75n2.

24. Gould, *Loch Ness Monster,* 1969, pp. 111n2, 112n1. See also Isobel Knight, "The Annals of the Loch Ness Monster," *Scotland's Magazine,* July 1957, pp. 40–42; Roy P. Mackal, *The Monsters of Loch Ness,* Chicago: Swallow, 1976, pp. 379–80.

25. W. H. Lehn, "Atmospheric Refraction and Lake Monsters," *Science,* 13 July 1979, pp. 183–85.

26. Gould, *Loch Ness Monster,* 1969, pp. 107, 110.

27. Dinsdale, *Loch Ness Monster,* 1961, pp. 81, 94–96, 110. See also "Our Special Artist Investigates the Loch Ness Monster," *Illustrated London News,* 13 Jan. 1934, pp. 1, 40–41.

28. For the Oxford and Cambridge expedition, see Maurice Burton, *The Elusive Monster,* London: Rupert Hart-Davis, 1961, p. 101. William Owen discusses his sightings in *Scotland's Loch Ness Monster,* Norwich (Eng.): Jarrold and Sons, 1980; and Mackal recounts how he was fooled in *Monsters of Loch Ness,* pp. 19, 83–84.

29. For a discussion of "Nessie weather," see Tim Dinsdale, *Monster Hunt,* Washington, D.C.: Acropolis, 1972, p. 195; Holiday, *Great Orm,* pp. 171–72; Mackal, *Monsters of Loch Ness,* p. 88. The conditions under which Fraser filmed the monster are recounted in Witchell, *Loch Ness Story,* 1974, p. 72. For the LNI sightings, see Mackal, *Monsters of Loch Ness,* p. 291.

30. Mackal, *Monsters of Loch Ness,* p. 88.

31. Whyte, *More than a Legend,* 1961, p. 110.

32. The figure 3,000 has been used by many people since its first appearance in print in 1961 (see David James, "Time to Meet the Monster," *Field,* 23 Nov. 1961, pp. 951–53), but no listing of all reported sightings was available until 1984. Ulrich Magin has now prepared a list with entries through 1984; he found only 600 reports, albeit the number of witnesses is a multiple of that figure. A list based on Magin's is given as appendix B. For estimates of mistaken reports, see Mackal, *Monsters of Loch Ness,* p. 84; Daniel Cohen, *A Modern Look at Monsters,* New York: Dodd, Mead, 1970, p. 96.

33. For the Campbell sighting, see Holiday, *Great Orm,* p. 65. The AAS photographs are discussed by Charles W. Wyckoff in "Loch Ness and Underwater Photography," *Technology Review,* Dec. 1976, p. 50.

34. The surgeon's photograph is discussed in Witchell, *Loch Ness Story,* 1974, p. 66; Ronald Binns, *The Loch Ness Mystery Solved,* Shepton Mallet (Somerset): Open Books, 1983, p. 96; Mackal, *Monsters of Loch Ness,* p. 96; "Je suis le père," *Match* (Paris), 11 Apr. 1959.

35. See Richard Whittington-Egan, "Loch Ness: The Monstrous Zoological Problem," *Contemporary Review,* Sept. 1974, pp. 138–44; Steuart Campbell, "The Surgeon's Monster Hoax," *British Journal of Photography,* 20 Apr. 1984, pp. 402–5, 410.

2

The Monster Exists

During the most recent of the glacial epochs Scotland was buried under ice. Immediately after the thaw, more than 10,000 years ago, the oceans covered a great deal of what is now land;[1] many of the present freshwater lochs were then fjords or saltwater lochs. Into these fjords came many species of marine life: some made a permanent home there, others were temporary visitors come to feed or to spawn. Similar conditions prevailed in other parts of the Northern Hemisphere—in Iceland, Scandinavia, Siberia, and North America.

The land, freed from the heavy burden of ice, rose very slowly, as it continues to do in many places, a few millimeters per year. Eventually some of the fjords were cut off from the oceans. The salts in the water became progressively more dilute as the lakes, as they now were, continually received fresh water as rain and run-off from hills and streams, and continually overflowed into the sea. In the end these erstwhile fjords became freshwater lakes.[2]

During these changes of land elevation and water level, many marine creatures were trapped in the fjords-become-lakes. Over the centuries and millennia, some of those species became extinct; others, adapting to the change from salt water to fresh, persisted and even flourished, albeit through altering in a number of ways their life-style, life cycle, even their physiology and anatomy. (In some parts of the world such typically marine creatures as sharks and porpoises have adapted themselves without visible change to life in wholly fresh water.[3])

In a number of the newly formed lakes of the Northern Hemisphere there had been trapped members of that rare, still-not-well-known sort of marine animal, the long-necked sea serpent.[4] These large fish-predators adapted well to the progressive lowering of the salinity of the water, but their continued existence depended on the presence of a large enough stock of fish to support a viable, breeding population—not less than twenty or thirty individuals, say. Hence, in many of the smaller lakes the sea serpents died out; they survive nowadays only in such very large lakes as Morar, Ness, and Shiel in Scotland, Okanagan and Champlain in North America, Lagerflot in Iceland, and possibly in a few others.

These northern lakes are remote from the more temperate regions in which Western civilization has developed, and it was well into the twentieth century before savants began to learn of the existence of the freshwater sea serpents. The indigenes, of course, knew them quite well as part of their accustomed surroundings. The Vikings had been very familiar with the saltwater sea serpents and had modeled the shapes of their vessels after the bulky bodies, long and tapering necks, and small heads,[5] reminiscent at the same time of such different creatures as horses, turtles, and snails: flat in the forehead; protruding but rather stubby jaws; eyes clearly visible only at some angles; and, near the top of the head, small projections whose nature remains unknown—horns, breathing tubes, ears? The Indians of North America knew these creatures in their lakes, as did the Celts and the Picts of Ireland and Scotland.[6]

Until the twentieth century almost all knowledge about these freshwater sea serpents came from the oral traditions of the natives. The animals were probably as rare a millennium or two ago as they are now; and their life-style being such that they rarely spend much time on the surface, they were and are seen quite infrequently. The little knowledge of them that came from direct observation was inevitably augmented by the imagination, so that the folklore became a confused mixture of fact and fancy, biology and myth. Some of the dragon tales borrowed from knowledge of sea serpents, as did the Scottish legends of water-kelpies, water-horses, and water-bulls. Even today one finds in the Highlands of Scotland an older generation to whom these creatures are both animal and omen: a mixture that is hardly understandable to one reared amid concrete and electricity and urban schooling, yet a mixture that was common to all peoples not so long ago and remains natural to many.

Scientists first learned of the freshwater sea serpents when the road to Inverness was rebuilt and made readily accessible to motor traffic in the early 1930s, and thus gave tourists a clear view over the waters of Loch Ness for some twenty miles—a view much better in 1933 and 1934 than at any time since, as the trees and shrubs have grown up again between the road and the water, the local council having decided that it was too expensive to keep cutting back the growth.[7] The men on the road crews no doubt scanned the loch frequently during pauses in their work; and it is also plausible that the rock-blasting disturbed the Nessies and brought them more often to the surface, their sensitivity to noise being a frequently noted characteristic.[8] So there was an unusually large number of sightings of Nessies in 1933 and 1934, and the outside world paid attention for the first time. The locals were a little surprised and amused at the fuss made by the tourists over their familiar water-horse, an-Niseag (rather well translated from the Gaelic as Nessie). Reporters came not only from England but from all over the world; expeditions came, scientists and laymen waxed passionate over what was true and what was false, circuses and zoos wanted specimens, hoaxers had a grand time of it.

Science has had a difficult time with Nessie. The important evidence has come in slowly: an indistinct, amateurish photo (fig. 1) taken with a box camera in November 1933, followed a month later by a short piece of movie film which showed little more than the water being disturbed by the passage, just below the surface, of a medium-sized Nessie.[9] On 19 April 1934 a rather good shot (fig. 2) of a neck and head was obtained, but only as a silhouette against the sun, with no details that would permit the creature to be properly classified.[10] (The morphology and habits of sea serpents are still so little known that even identification as mammal or reptile remains problematic.) Most of the information about Nessies has come from human observers and local folklore, of no great help to biologists; and so, to the present, only a handful of zoologists have spent much time on the matter. As with any problem of this sort, where no viable way of solving it is apparent, most scientists pay little attention and some say the problem does not exist.

So progress has come slowly, through the work of dedicated amateurs. Eyewitness reports and the few photographs, obtained largely by chance, have been gathered in the classic books *The Loch Ness Monster and Others* (1934), by Rupert Gould, and *More than a Legend* (1957), by Constance Whyte. Gould's description, culled from more than fifty eyewitnesses and two photographs, has been confirmed by later work: Nessies, he said, grow to some forty-five feet (perhaps ten feet of neck, twenty feet of body, and the rest tail), with a maximum diameter of about five

Figure 1. The first photograph ever purportedly of the Loch Ness monster, taken on 12 November 1933 by Hugh Gray. This print is from a glass lantern–slide (a contact positive) made in 1933 for E. Heron-Allen from the original Gray negative; two such slides came into the possession of Maurice Burton twenty years ago, and Steuart Campbell was instrumental in obtaining one of them for the Fortean Picture Library. The slide makes possible a far higher quality of reproduction than the glossy prints available from the newspaper files. Reproduced courtesy of Fortean Picture Library.

Figure 2. *Top:* The most famous Nessie photograph, referred to as the "surgeon's" photograph, taken by R. K. Wilson in April 1934. Reproduced courtesy of the *London Daily Mail. Bottom:* A little-known second shot, taken by Wilson on the same occasion.

Figure 3. A photograph said to be of Nessie, taken by Lachlan Stuart on 14 July 1951. Reproduced courtesy of the London Express News and Feature Services.

Figure 4. The Macnab photograph, taken on 29 July 1955. Reproduced courtesy of Camera Press Ltd.

feet; they have small heads and dark skin that lightens in sun or air—a rough, granulated or warty skin, with a ridge along the crest of the back; the body can display a variable number of humps of variable shape, and the paddles or flippers are placed quite low on the body.[11] Two aspects of the humped appearance are confirmed by Macnab's photo (fig. 4), which was published by Whyte, and by Dinsdale's film (fig. 5).

Gould noted references to these creatures in the seventh-century biography of St. Columba and in folklore attaching to several other lochs; he also obtained, at first- or secondhand, reports of sightings from 1871, the 1880s, 1895, 1903, 1914–18, 1923, 1929, 1930, and 1932—demonstrating that Nessies were known locally long before the outer world became fascinated by them in 1933. Whyte has offered a discussion full of insight into the mythical components that accrued in the traditional stories, and she gathered scores of reports of sightings from people she knew (having lived for twenty years at Clachnaharry in Inverness). She gave similar information about the related creatures in Shiel and Morar, as well as snippets of such reports from Iceland and British Columbia.[12]

Dinsdale's 1960 film (fig. 5; see also fig. 6) substantiates the notable speeds of which Nessies are capable (ten miles per hour at the least), as does the sonar work of the team from Birmingham University in 1968. The Loch Ness Investigation Bureau (LNI) has added a few pieces of film that seem to confirm that Nessies are fish-predators (salmon are seen jumping with a Nessie in pursuit) and occasionally amphibious (a medium-sized Nessie was seen at least partly out of the water on a beach, unfortunately at two-miles range).[13]

The Academy of Applied Science (AAS) has obtained the best details yet: in 1972 a diamond-shaped flipper about six feet long and two feet wide was captured at two different angles in underwater photographs taken forty-five seconds apart (fig. 7). And in 1975 the Academy had real luck. One shot (fig. 8) captured the front of a Nessie, showing parts of two flippers and the long neck, a total length of about seventeen feet. Another shot (fig. 9) was claimed to be of the head, indicating the presence of the horns or stalks occasionally mentioned by eyewitnesses, but showing little other detail.[14] Although classification of the creatures was still not possible, it did seem time to assign a formal name, to permit

Figure 5. Stills from Tim Dinsdale's film of 1960, enlarged from original 16mm Kodak Plus X reversal film. Reproduced courtesy of Tim Dinsdale and UPITN Film Library; captions provided by Tim Dinsdale.

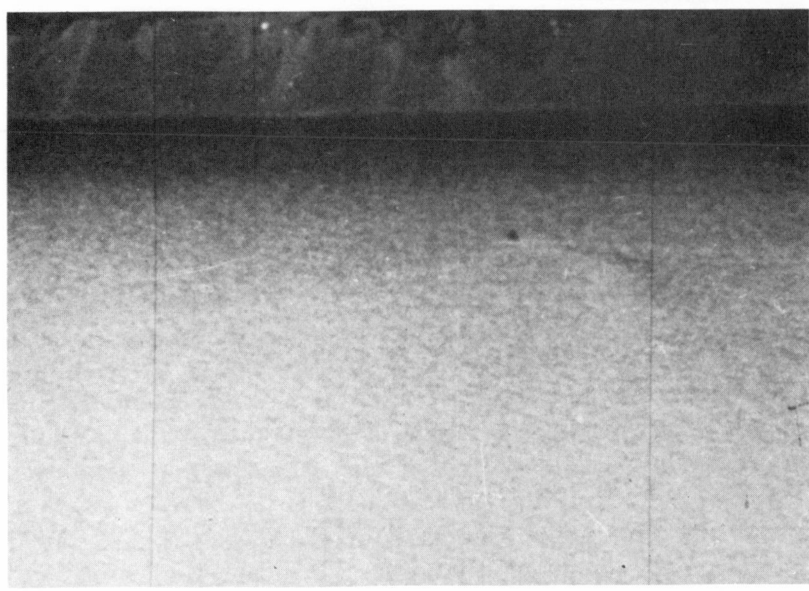

(A) Hump first seen at approximately 1,650 yards, just starting to move to left.

(B) Zig-zagging slowly to the right, glassy wake developing (1966 Royal Air Force report estimates width of hump at six feet and height at three feet, neither a surface boat nor submarine, moves at up to ten miles per hour, is "probably animate").

(C) Momentarily a second hump appears behind the first (see fig. 6).

(D) Humps beginning to submerge as the creature approaches the far shore (note seagull in flight), then fully submerge.

(E) Turns abruptly left at right angles.

(F) Swims subsurface, throwing up a big wash (note vehicle driving east along north-shore road, which is identifiable—it really *is* Loch Ness).

(G) Compare shot of fourteen-foot heavy clinker boat at approximately the same range as the hump in (A), powered by a five-horsepower outboard, doing seven miles per hour.

(H) Boat travels west by far shore at full throttle; compare with monster's wake in (F).

(I) Boat at approximately 800 yards, off Foyers point.

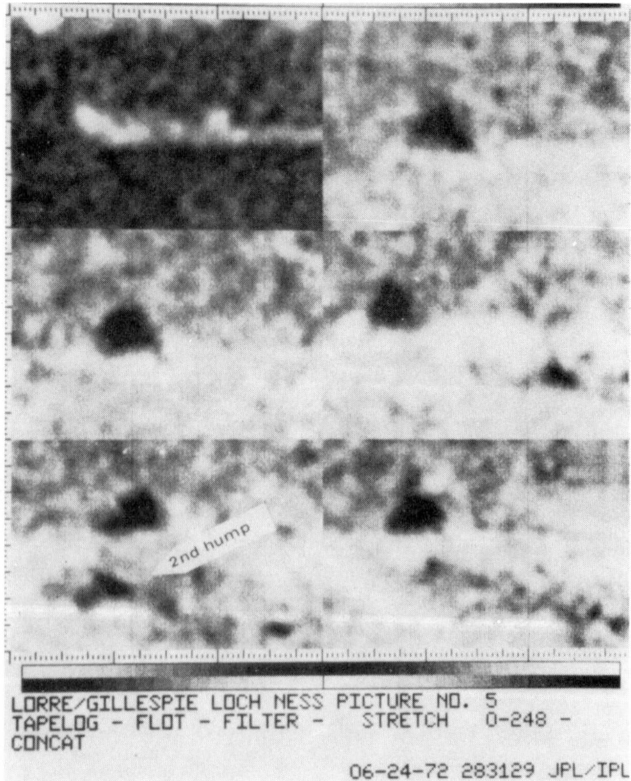

Figure 6. Computer-enhancement of Tim Dinsdale's 1960 film confirms indications of a second hump following the larger one. Reproduced courtesy of the Academy of Applied Science.

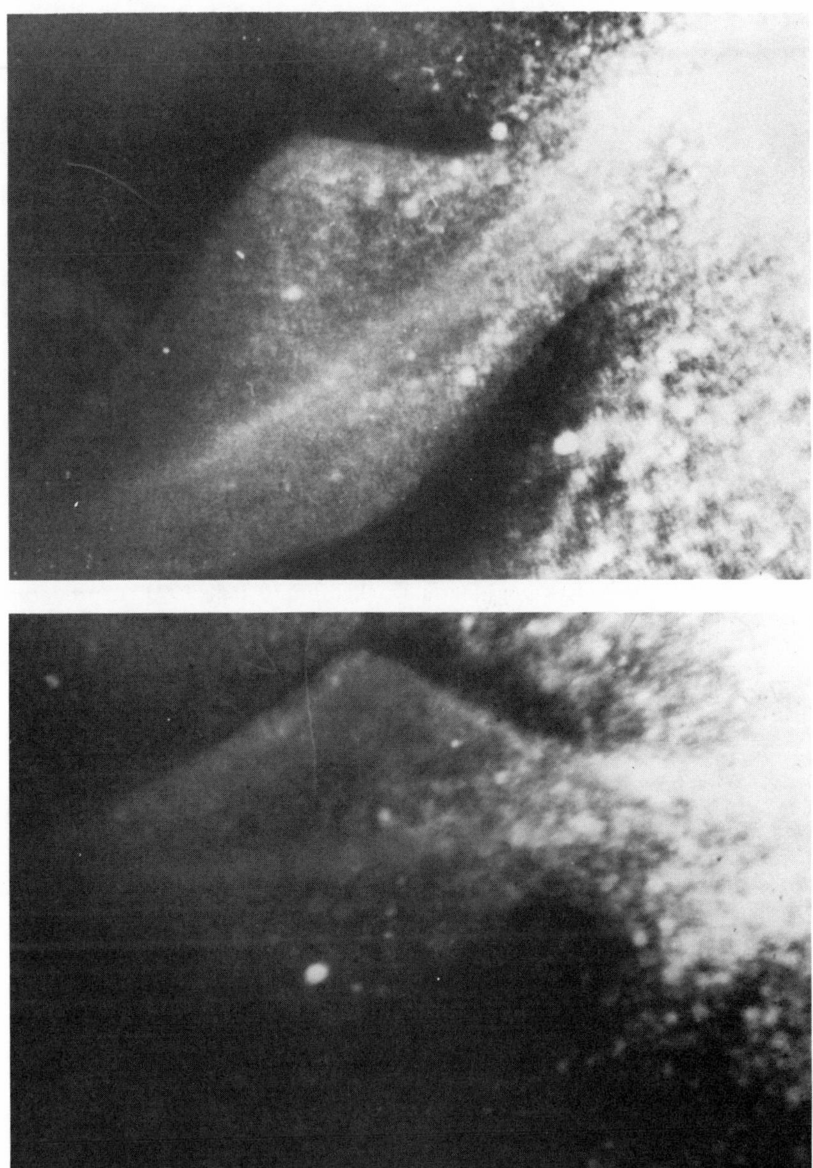

Figure 7. The "flipper" photographs, taken on 8 August 1972. Reproduced courtesy of the Academy of Applied Science.

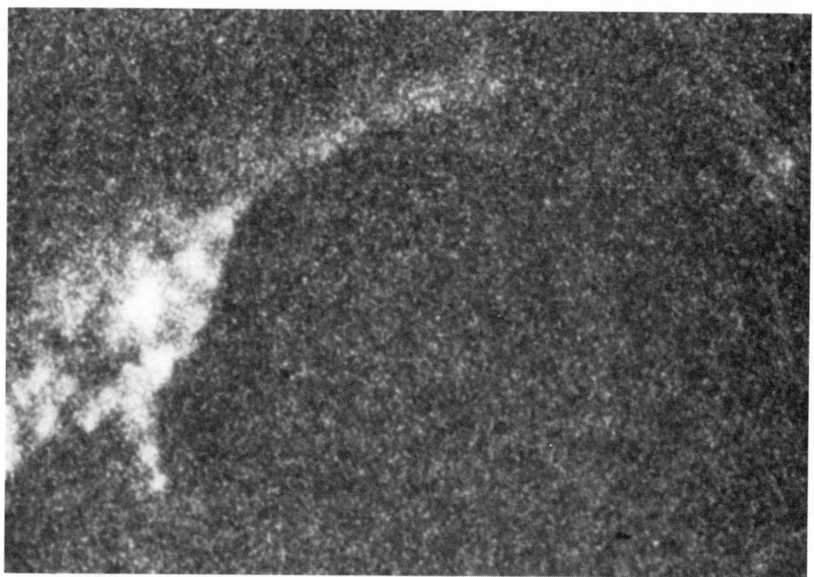

Figure 8. The "body-neck" photograph, taken on 20 June 1975. Reproduced courtesy of the Academy of Applied Science.

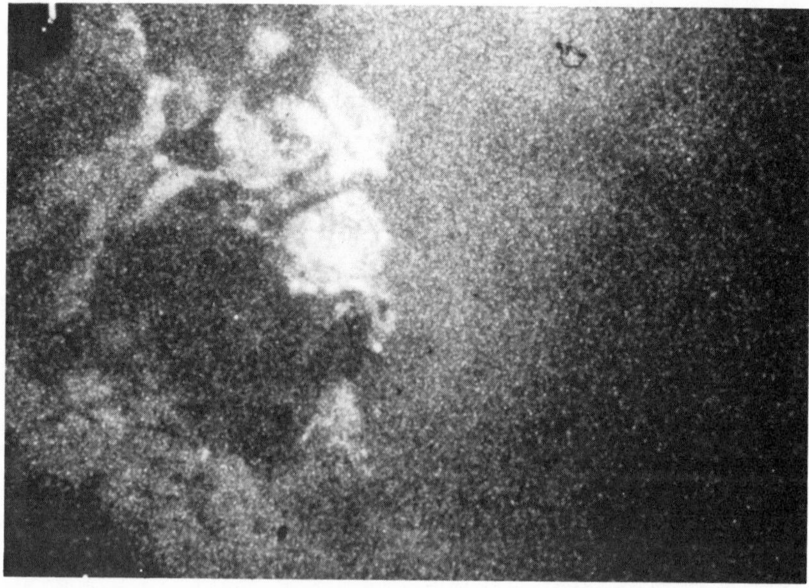

Figure 9. The "gargoyle head" photograph, taken on 20 June 1975. Reproduced courtesy of the Academy of Applied Science.

official protection to be extended to the species. Scott and Rines perforce used descriptive information only as they gave to the Loch Ness monster, or Nessie, the designation *Nessiteras rhombopteryx*—wonder of Ness with the diamond-shaped fin.[15]

Each year since 1975 has seen expeditions seeking better photographs, probing with sonar, and searching for bones and carcasses, but the results of that year have so far remained unmatched. The Loch Ness and Morar Project obtained dozens of strong sonar echoes in 1982 and is seeking conclusive identification of the objects responsible for the soundings. Most exciting and most promising are the plans of the AAS to use dolphins in the search; fitted with cameras and strobes actuated by sonar, they will be trained to seek out large creatures.[16]

Notes

1. Andrew Goudie, *Environmental Change,* Oxford: Clarendon Press, 1977; Kalerno Rankama, ed., *The Quaternary,* vol. 2, New York: Interscience, 1967.

2. Tim Dinsdale, *The Loch Ness Monster,* London: Routledge and Kegan Paul, 1961, p. 66; Constance Whyte, *More than a Legend,* rev. 3d imp., London: Hamish Hamilton, 1961, pp. 23, 149, 165; Nicholas Witchell, *The Loch Ness Story,* Lavenham (Suffolk): Terence Dalton, 1974, pp. 17, 19, 201.

3. Tim Dinsdale, *The Story of the Loch Ness Monster,* London: Allan Wingate and Target (Universal Tandem), 1973, p. 56.

4. Rupert T. Gould, *The Case for the Sea-Serpent,* London: Philip Allan, 1930; Bernard Heuvelmans, *In the Wake of the Sea Serpents,* New York: Hill and Wang, 1968.

5. Tim Dinsdale, *Project Water Horse,* London: Routledge and Kegan Paul, 1975, p. 158; Elliott Snow, "The Great Sea Serpent," *Nature Magazine,* Oct. 1927, pp. 240–43.

6. Whyte, *More than a Legend,* 1961, p. 63, 123–33, 135, 136, 140–45, 146–51, 158–60.

7. Maurice Burton, *The Elusive Monster,* London: Rupert Hart-Davis, 1961, p. 13; Dinsdale, *Loch Ness Monster,* 1961, p. 14; Roy P. Mackal, *The Monsters of Loch Ness,* Chicago: Swallow, 1976, p. 85; Whyte, *More than a Legend,* 1961, pp. 21–22, 31.

8. Mackal, *Monsters of Loch Ness,* p. 85; Rupert T. Gould, *The Loch Ness Monster and Others,* New York: University Books, 1969, pp. 82, 84; F. W. Holiday, *The Great Orm of Loch Ness,* New York: Avon, 1970, pp. 89, 92, 172; Witchell, *Loch Ness Story,* 1974, p. 105.

9. For the Hugh Gray photograph, see Dinsdale, *Loch Ness Monster,*

1961, pp. 43, 88; Gould, *Loch Ness Monster,* 1969, p. 23; Whyte, *More than a Legend,* 1961, pp. 2–4. For the movie film, see Burton, *Elusive Monster,* p. 72; Mackal, *Monsters of Loch Ness,* p. 290.

10. See Steuart Campbell, "The Surgeon's Monster Hoax," *British Journal of Photography,* 20 Apr. 1984, pp. 402–5, 410; Dinsdale, *Loch Ness Monster,* 1961, pp. 46, 68–75; Gould, *Loch Ness Monster,* 1969, pp. 23–24; Whyte, *More than a Legend,* 1961, pp. 6–8; Witchell, *Loch Ness Story,* 1974, pp. 66–69.

11. Gould, *Loch Ness Monster,* 1969, pp. 157–58.

12. Whyte discussed the mythical component in *More than a Legend,* 1961, pp. 63–64, 135, 140–45, 158–60. For her account of related creatures, see ibid., pp. 123–33, 136, 146–51.

13. For his discussion of the 1960 film, see Dinsdale, *Loch Ness Monster,* 1961, p. 6; ibid., 1972, p. 125; ibid., 1982. The sonar work is recounted in Hugh Braithwaite, "Sonar Picks Up Stirrings in Loch Ness," *New Scientist,* 19 Dec. 1968, pp. 664–66. Mackal, in *Monsters of Loch Ness,* pp. 290–94, discusses the LNI confirmation.

14. For further discussion of the "flipper," "body-neck," and "gargoyle head" photographs, see Robert H. Rines et al., "Search for the Loch Ness Monster," *Technology Review,* Mar.–Apr. 1976, pp. 25–40; Nigel Sitwell, "The Loch Ness Monster Evidence," *Wildlife,* Mar. 1976, pp. 102–9; "Skeptical Eye—The (Retouched) Loch Ness Monster," *Discover,* Sept. 1984, p. 6; Charles W. Wyckoff, unpublished letter to Henry Anatole Grunwald, ed., *Discover,* dated 27 Aug. 1984. See also p. 72, n. 5.

15. "Naming the Loch Ness Monster," *Nature,* 11 Dec. 1975, pp. 466–68.

16. Loch Ness and Morar Project, *Report—1983.* The AAS plans are discussed in Barry Fox, "Patents-Camera for Dolphins," *New Scientist,* 17 June 1982, p. 779; "Dolphin Chase for Nessie," *New Scientist,* 21 June 1979, p. 982; "Ness Hunter Dies," *New Scientist,* 5 July 1979, p. 6; "Dolphins to Look for Nessie," *Science,* 13 Apr. 1979, p. 159; Nigel Sitwell, "Loch Ness Dolphins," *New Scientist,* 1 July 1982, p. 47.

3

Introduction Proper

I have not, in the preceding pages, tampered significantly with any "facts." The references given are genuine and can be easily checked; I have "Lester Smith's" letters in my files, which will be kept permanently in my university's archives. The mutually contradictory explanations of the Loch Ness phenomenon in chapters 1 and 2 do not differ so much on the facts as on the interpretation of those facts; in that sense, both explanations are truthful—as far as they go. The *whole* truth, however, is something else again. This book is an attempt to move in that direction.

There are many other topics besides Loch Ness that arouse great public interest and that could be introduced in a similar manner. Fringe subjects, one might call them, or anomalies. They find themselves in a no-man's-land between the mainstream of science on the one side and the quicksands of pseudoscience on the other: neither accepted and reliable like the paradigms of chemistry, say, nor obviously and demonstrably absurd like the notion that the earth is flat.

Although there is general agreement that such a class of subjects exists, there are arguments about the inclusion in that class of any given item. Those who accept the interpretations in chapter 2, for example, insist that Nessies are properly a subject for science; whereas those who are convinced by the interpretations in chapter 1 know the Loch Ness monster to be pseudoscience. For me, the existence of a significant division of opinion about such a topic identifies it as a fringe subject. On that criterion my inventory of the fringes is huge: acupuncture, ancient astronauts, astrology, Bermuda Triangle, dowsing, extrasensory perception (also psi, parapsychology, precognition, psychokinesis, telepathy *et alia*); Fortean themes, Kirlian photography, homeopathy, and laetrile and other unconventional curatives; phrenology, Sasquatch (abominable snowman, Bigfoot, yeti), UFOs, Velikovsky—and much more besides.

Consistent with the criterion that disagreement exists, fringe subjects are controversial. It is my contention that these controversies have a number of features in common and that an understanding of these commonalities is worth striving for: it permits one to think about any of these topics in a more efficient manner, and it has as an inevitable corollary an appreciation of what science is and what it is not. My aim in this book is to reach for that sort of understanding through a detailed explication of the controversy about the Loch Ness monster. The Loch Ness affair well illustrates, I believe, some general and important aspects of the interaction of science with the wider society; for example, that science generally dismisses (at first, at least) claims by laymen of unusual events, phenomena, or theories; that outsiders can rarely induce scientists to take such things seriously; and that the interested observer finds it difficult to make sense of the ensuing argument and to reach a reasonable judgment.

The most persistent searcher for Nessie, Tim Dinsdale, has suggested (in *Monster Hunt*, a revised edition of his earlier work *The Leviathans*) that the Loch Ness animals themselves are not the primary issue. At stake at Loch Ness, he maintains, is the matter of truth and how human beings approach that, with science being one of those approaches—and the most successful to date. I agree and seek to point to general truths that are illustrated by this marvelous, confused, long-drawn-out controversy: truths that are largely independent of the actual existence of Loch Ness monsters.

The enigma of Loch Ness arose inevitably, I believe, from certain social realities: the nature of scientific activity and its place in society, the roles of newspapers and magazines and broadcasting, the mind-set of our culture. But in analyzing the enigma, perforce I make statements that might seem to leave few of the actors untouched by criticism. I ask those mentioned, and my readers, to recognize that criticisms of actions are not synonymously sweeping condemnations of the actors: it is the essence of the human condition to make mistakes, and social practices do not allow any of us the luxury of pursuing even the most noble ends by unsullied noble means. I respect the achievements of biological science and those who practice it; I respect the monster hunters and their endeavor; and I ask all for their understanding as I draw attention to Achilles' heels. I hope that biologists and hunters alike will, as I do, see in this book a confirmation of the wider significance of their own labors, a seeking of social truth in company with their seeking of truth about the world we inhabit.

4

The Conventional Wisdom

It is commonly said that evidence speaks for itself. Even a moment's reflection, however, reveals that to be untrue. Evidence can "speak" only if an interest is taken in it by someone; and then it speaks only to that individual, and only in a manner that he is prepared to hear.

We all form opinions about all sorts of issues, not by examining the evidence ourselves, but by adopting at second- or thirdhand the opinions of what we take to be appropriate authorities. And, especially when there is a range of expert opinions, we select or even transform those opinions as we filter them through our personal sets of preconceptions about society, science, morality, and the rest. Such an issue as the possible existence of Nessies we do not pursue by searching libraries, by visiting Loch Ness, by talking with those who hunt Nessies. Since this is a matter of zoology, we look to the zoologists for an answer. And we do not even seek a direct answer: we take one indirectly, from the popularizers of science and from magazines and newspapers. Those secondary sources, of course, tend to take their answers and attitudes from those whose profession is biology. So the conventional wisdom is unlikely to include the existence of Nessies until there is a clear consensus to that effect among scientists in the appropriate field, and that cannot happen unless the matter appears to them to be of scientific interest. In chapter 7 I explain why the zoologists on the whole have not yet found Nessies to be a subject worthy of their time.

Despite this scientific consensus, the conventional wisdom of the wider society could still reflect a strong disposition to believe in Nessies if the media of popular communication took such a stance. The media are influenced by many considerations, of which the straightforward presentation of the best existing knowledge is not necessarily the most decisive. A scientific consensus is ignored if the social or political pressures are sufficiently strong, not an infrequent situation in economics, for example. That is less likely to arise in matters of biology, though there are clear instances at hand there also—on the issue of hereditary correlates of ability, for example; or that of the teaching of "scientific creationism" together with the theory of evolution; or that of abortion and when, after conception, a human life meaningfully begins. But the existence of Nessies, or their failure to exist, has no obvious social or political implications that seem important to most people, and the attitudes of the media have not been constrained thereby. So the popularly conveyed conventional wisdom has largely followed the scientific consensus that formed in the years 1933 and 1934. A journalist or editor tempted to question that traditional view would soon find good reasons to resist the temptation: in chapters 5 and 6 I indicate why one might well find cause to disbelieve the monster hunters; to dismiss the evidence wholesale, particularly in view of the perpetration of undoubted hoaxes and frauds; to class Nessies with such intellectually discredited subjects as astrology.

The Loch Ness monster is nowadays a cliché. Any serpentine shape with a suggestion of tartan or whiskey needs no further introduction or explication, as every cartoonist knows. A breakfast cereal is commended in a television commercial by kilted oarsmen whose boat is lifted out of the water by a huge leathery back.[1] The fearsome roller-coaster at Busch Gardens (in Williamsburg, Virginia) is named the Loch Ness Monster. Even as I write this, I receive a mail-order catalog of "unique gifts from the realm of science" that offers a Plesiosaurus Fossil Kit whose description includes, "'Nessy' [*sic*] is said to be a survivor of this creature."[2] Nessie is a cliché, as well known as the Great Depression, the South Sea Bubble, or the unicorn. As with all clichés the connotation is simplistic in the extreme—a headline, not a nuanced description or story; and the connotation for Nessie is myth, gullibility, joke.

Only 13 percent of Americans in 1978 gave credence to the evidence that Nessie exists, roughly the same as for Bigfoot and for ghosts and much less than for UFOs (57 percent) or telepathy (51 percent), for example.[3] Teenagers were more inclined to believe: in 1978, 31 percent gave credence to Nessie, 40 percent to Bigfoot, and 67 percent to extrasensory perception; but by 1984 the percentages of believers had declined to 18, 24, and 59 respectively.[4]

In seeking the genesis of the public image, I consider first the period before Nessie caused a public fuss in 1933.

PREHISTORY: BEFORE 1933

Nineteen thirty-three was not a good year for enlisting biologists in a search for a quite improbably large animal in a Scottish freshwater loch, an animal quite improbable in shape and in behavior as well as in size. Western society was economically blighted, socially fragmented, in political turmoil; radicals, conservatives, and those in between were agreed that the very survival of societies and nations was in question. I have no doubt that some Marxist commentators, with their unfailingly impeccable logic, diagnosed the worldwide publicity about the Loch Ness monster as a circus provided by the capitalist press to distract the proletariat from its revolutionary duty and destiny. Indeed, in 1968 a Soviet scientist was reported to say: "When it becomes necessary to distract the reader from tough problems, Western leaders have three ready sensations which never fail—Flying Saucers, the Loch Ness Monster and the Abominable Snowman."[5] (That these subjects are referred to also in Soviet publications no doubt means that Soviet leaders, too, on occasion wish to distract their populace from certain tough problems.)

Perhaps it could have been different. Had Nazism attacked the Loch Ness monster as a figment of Jewish science (as, for example, it attacked modern physics), free scientists might have been driven to defend at least the possibility of Nessies. But that is the sheerest wishful thinking. It seems not unreasonable to suggest, however, that the scientific community was as anxious as the larger society to maintain the security and the desirable qualities and values and substance of the earlier status quo that seemed in danger of being lost. Perhaps most clearly in the Nazi movement one could see a discarding of traditional views and values, a rejection of the intellect in favor of unbridled emotions, a discounting of the truths of science and technology in a romantic endeavor to harness to political ends merely the tangible fruits and not the insights brought by those endeavors. Certainly most clearly in physics one saw the Nazis saying that the work of science must be governed by theories based on political philosophy, not theories generated by science itself. Under those circumstances free scientists would have been doubly anxious about the security of their freedom to pursue their work, apprehensive of any danger that their work might be dictated by the uninformed and the unlearned, were they in government, in capitalist business, in the trade unions, among the laity, or in the popular press.

Within science itself there had also been turmoil. Confidence in the possibility of achieving total understanding of physical reality through quantitative application of commonsense notions had been high before the turn of the century. Physics had a manageably small number of fundamental particles behaving themselves under the macroscopic laws of energy and mechanics; chemistry had a nicely categorized set of indestructible atoms, whose combination in simple ways explained the existence and behavior of all materials. Biology had simple, quantitative rules of heredity and an overarching explanation of the evolution of the fittest and most complex in easy stages from the most primitive and simple.

All that changed over the ensuing three decades. Particles were no longer just particles but sometimes behaved as insubstantial waves—even in the absence of a medium in which to propagate themselves. Newton's simple, commonsense mechanics had been superseded by a relativity theory that postulated ludicrously paradoxical possibilities (that two identical twins would age at different rates if they were exposed to such different environments as motion at very high speeds, for example—a space traveler might return middle-aged, to find his twin dead long ago of old age). Atoms were not indestructible: they changed into atoms of other kinds, as though the alchemists had been right after all. And biologists began to doubt that matters of heredity and evolution were quite as straightforward as had been thought previously.

There was pronounced resistance by some of the older generation against the new approaches, theories, and discoveries. Relativity was strongly resisted, as were at least certain implications and interpretations of the new quantum physics or wave mechanics. Indeed, the Nazi attack on "Jewish" science was instigated by some physicists of the old school: unable to win their battle by reason and evidence within their discipline, they enlisted the aid of government, seizing on the happenstance that some of the best physicists of modern view were Jews.[6]

In the wider society, then, and also in the scientific community, traditional approaches were under attack, and lines were drawn between the new, the radical, the young and tradition and its guardians. A pronounced division of that sort is not always with us, though we speak of the inevitability of the generation gap. There have been periods, and quite lengthy ones, during which such differences have been minor in comparison to the strength of the societal consensus on fundamentals: for long centuries in Japan, in Europe, in China. In matters of science, the confident march forward on all fronts during the nineteenth century was largely cohesive, as new discoveries and ideas fitted readily with what was already known, and the young and old generations shared a common approach. They may have differed in status and influence, but they did not subscribe to different worldviews.

So it is not trivial, I maintain, to note the clashes of philosophies that occurred more or less violently in the years and decades preceding 1933. Perhaps the existence of Nessies, confounding the Establishment, would have been welcomed by some; but surely there existed also, within the Establishment, a greater resistance than "usual" to radically, startlingly new claims in any field, including biology, especially when the claims came from outsiders, from nonscientists. Had Nessies showed themselves to many in the middle of the nineteenth century, I suspect that they would have been sought with relish by biologists and used pro and con in the debate over Darwin's ideas. But 1933 was not a good year.

There is also a quite specific reason to be found in these "prehistorical" years for an attitude of skepticism toward Nessies. The oceanic cousin of Nessie, the "great sea serpent," had become persona non grata to science even before the turn of the century; the reasons for that are another story, but the fact is clear. In the 1890s sea serpents were a stock subject for jesting.[7] A few zoologists remained open-minded or even expressed belief in their reality, but they were a very small minority.[8] Such creatures were rejected by science and treated by the press and public as folklore and myth, to the extent that those who reported seeing them were mercilessly ridiculed; as a result, many observers kept their experiences to themselves. I include here a few of the frequently quoted illustrations of that.

Colonel Perkins, of Boston, in about 1850 described himself as "one of the unfortunate individuals who saw it himself"; Captain Dewar, some twenty-five years later, recounted that his "relations wrote saying that they would have seen a hundred sea serpents and never reported it, and a lady wrote saying she pitied anyone who was related to anyone who had seen a sea serpent."[9] Vice-Admiral H. L. Fleet wrote, thirty years after the event, of having seen what he and a colleague took to be a sea serpent, "but decided to say nothing about it, having due regard to the scepticism of the British public"; and Rupert Gould knew personally five naval officers who kept similar sightings to themselves.[10] Heuvelmans also has personal knowledge of the suppression of news of sightings for fear of ridicule.[11] Captain R. J. Cringle of the *Umfuli,* from which a sea serpent was seen in 1893, wrote in 1929: "I have suffered so much ridicule on this that I must decline to have anything more to do with it"; and "I have many times wished that anybody else had seen that sea-monster rather than me."[12]

Nor has the situation changed in any obvious manner up to the present. A few biologists helped to found the International Society of Cryptozoology, which takes seriously the search for unexpected animals; but *Nature,* for example, has ignored sea serpents for nigh on a century, and such creatures continue to be not uncommonly used in cartoons and jokes.[13]

1933: LOCH NESS BECOMES NEWS

Among those who lived around Loch Ness, stories about large water animals were common long before 1933. As I have indicated in chapters 1 and 2, for some people these were real animals and quite a natural sort of creature, whereas for others they were somewhat mysterious and semimythic; for some others again they were solely legendary. Oral reports from the nineteenth century, at second- and thirdhand, were obtained by Gould, but reliable written records of actual sightings in Loch Ness seem to date only from 1930, when the *Northern Chronicle* published the report of a sighting and related correspondence.[14]

The widespread publicity began with a report in the *Inverness Courier* on 2 May 1933. During the next five or six months there were at least thirty purported sightings and perhaps the same number of news items about them, largely in the local papers only.[15] This succession of reported sightings eventually attracted national and then international attention, and by November many reporters from far-off papers were at the scene. Contemporaneous accounts and recollections of those involved indicate rather conclusively that the big fuss began no earlier than October;[16] the first articles in the *Scotsman* appeared in mid-October and in the *Times* (London) only in December.

Experts surmised that some marine creature had somehow gotten into Loch Ness—a large seal, or a beluga (white whale), or a shark; there were also some more exotic suggestions. But the actual descriptions did not really fit any known creature, and some commentators remarked that gullible souls were again misinterpreting some vaguely seen phenomenon in terms of that hoary superstition, the great sea serpent, albeit this time in a lake rather than at sea. Such a juxtaposition became firmer when Rupert Gould said plainly that indeed a sea serpent had somehow entered Loch Ness.[17] Several years earlier Gould had published *The Case for the Sea-Serpent* in which he maintained that, contrary to the received wisdom in biology, sea serpents were a reality. By now connecting the Loch Ness monster with the sea serpent, Gould inevitably, though unintentionally, ensured that Nessie would be included in the rejection by established science of the existence of such creatures.

This rejection was undoubtedly assisted by the improbable nature of several reports that the monster had also been seen on land (see chap. 5), and by the perpetration, as early as December 1933 and January 1934, of several hoaxes (see chap. 6). The *Daily Mail* lost interest when it was the victim of a hoax. The *Times* (London), which gave considerable space to news items and letters almost daily for five or six weeks, decided in mid-January that the subject was no longer fit for serious discussion.[18] It mentioned the matter fewer than twenty times during the remainder of 1934, as against nearly fifty times between 8 December 1933 and 18 January 1934; and during the next twenty-five years it found only a half-dozen occasions to mention the Loch Ness monster.

So the initial skepticism of the experts, together with improbable claims by witnesses and the occurrence of hoaxes and the connection with sea serpents, quite quickly persuaded the newspapers that Nessie was nothing but a silly-season phenomenon. The media thereafter, and until the 1950s, mentioned the matter only rarely, and then in a tone of jocularity and disbelief. Inevitably that formed the public image of the Loch Ness monster: a cliché denoting gullibility and summer-season funny stories.

As with sea serpents, witnesses were reluctant to report sightings and expose themselves to the resulting ridicule. Westrum has pointed out that the failure to report sightings leads to an unwarranted belief among the public that sightings do not occur, and that in turn makes it even less likely that witnesses will feel free to report sightings.[19] In the present case we now have a continuing record of sightings for every year from 1933 to date,[20] but most of these were recorded or published only long after the sightings themselves.

REOPENING THE CASE: CONSTANCE WHYTE

Constance Whyte published an article in 1950 about the Loch Ness monster;[21] seven years later her book *More than a Legend* appeared. She had lived near Loch Ness for many years, had heard firsthand from many people who had seen the monster, and had become sufficiently interested to study many aspects of the matter. Her work produced a resurgence of interest to which can be traced the later expeditions and the burgeoning of articles and books on the subject. Moreover, the tone of what was written changed from the jocular disbelief of the 1930s and 1940s to a willingness to believe in the possible reality of the animals. I have read Whyte's book many times and continue to marvel at her marshaling of the evidence and her analysis of it: a veritable tour de force of history, eyewitness accounts, biology and geology, folklore, and perceptive commentary on the formation of belief, the motives and practices of newspapers and of experts, and much else.

Although the articles in magazines and the books about the

subject have increasingly taken a believing stance, the daily press has hardly altered its attitude of disbelief and jocularity, and the public image of Nessie remains much as it was in the 1930s. Public statements by British zoologists continue to deny that conclusive evidence is in hand, though some American experts have been cautiously positive even in public. An anonymously conducted poll of marine biologists has revealed that as many as 40 percent give credence to the reality of the animals.[22]

Full discussions are available elsewhere of the accumulation of evidence over the last twenty years and of the nature of the writings about the matter;[23] a synopsis appears as appendix A. For the present purpose it is enough to note that Nessie's public image is not yet that of a real animal species, rather still that of a cliché that can be relied on to induce smiles or laughter. In subsequent chapters I attempt to explicate the many reasons why no decisive change in that public image has occurred and why the Loch Ness monster remains outside the mainstream of science.

Notes

1. The 1981 commercial featured Kellogg's "Raisins, Rice, and Rye" cereal.

2. The catalog was from the Carolina Biological Supply Co., printed in 1981.

3. J. Richard Greenwell, "Academia and the Occult: An Experience at Arizona," *Skeptical Inquirer,* Fall 1980, pp. 39–46; J. Richard Greenwell and James E. King, "Scientists and Anomalous Phenomena: Preliminary Results of a Survey," *Zetetic Scholar,* July 1980, pp. 17–29.

4. George Gallup, "Gallup Youth Survey," 6 June 1984.

5. Quoted in Nicholas Witchell, *The Loch Ness Story,* Lavenham (Suffolk): Terence Dalton, 1974, p. 16.

6. See Alan D. Beyerchen, *Scientists under Hitler,* New Haven, Conn.: Yale University Press, 1977.

7. "What the Sea-Serpent Should Look Like," *Spectator,* 29 Sept. 1894, p. 400.

8. On the issue of zoologists' beliefs in sea serpents, see Rupert T. Gould, *The Case for the Sea-Serpent,* London: Philip Allan, 1930, pp. vii, 1; Bernard Heuvelmans, *In the Wake of the Sea Serpents,* New York: Hill and Wang, 1968, pp. 23–28; Constance Whyte, *More than a Legend,* rev. 3d imp., London: Hamish Hamilton, 1961, p. 151.

9. Quoted in Whyte, *More than a Legend,* 1961, p. 157.

10. Gould, *Case for the Sea-Serpent,* p. 189.

11. Heuvelmans, *Wake of the Sea Serpents*, p. 461.

12. Quoted in Gould, *Case for the Sea-Serpent*, pp. 188, 193.

13. See Ron Westrum, "Knowledge about Sea-Serpents," in Roy Wallis, ed., *Sociological Review*, monograph no. 27, Mar. 1979, pp. 293–314; Bernard Dixon, "Sea Serpent Survey," *Omni*, Sept. 1980, p. 18.

14. *Northern Chronicle*, 27 Aug. 1930; letters from "Camper," "Not an Angler," and "Invernessian," *Northern Chronicle*, 3 Sept. 1930; letters from "R.A.M.," and "Another Angler," *Northern Chronicle*, 10 Sept. 1930. See also letter from "Piscator," *Inverness Courier*, 29 Aug. 1930. The oral reports obtained by Rupert Gould are in *The Loch Ness Monster and Others*, New York: University Books, 1969.

15. On the purported sightings between May and November, see Henry Bauer, "The Loch Ness Monster: Public Perception and the Evidence," *Cryptozoology*, (no. 1, 1982): 40–45. For contemporary accounts, see the *Daily Express*, 1933, issues for 9, 28 June; 12 Aug.; 2 Sept.; *Inverness Courier*, 1933, issues for 2, 7, 9, 25 June; 4, 9 Aug.; 15 Sept.; *Northern Chronicle*, 1933, issues for 3, 8 May; 7, 14 June; 9, 12 Aug.

16. See Bauer in *Cryptozoology*, pp. 40–45; W. P. Pycraft, "Loch Ness in Possession of a 'Sea-Serpent'!" *Illustrated London News*, 11 Nov. 1933, pp. 760–61.

17. Rupert T. Gould, "The Loch Ness 'Monster'—A Survey of the Evidence—Fifty-one Witnesses," *Times* (London), 9 Dec. 1933, pp. 13–14.

18. See Bauer in *Cryptozoology*, pp. 40–45; Henry Bauer, "Society and Scientific Anomalies: Common Knowledge about the Loch Ness Monster," *Journal of Scientific Exploration*, 1(1987): 51–74; Henry Bauer, "Public Perception of the Loch Ness Monster," *Scottish Naturalist* (Proceedings of the International Society of Cryptozoology, Edinburgh, July 1987), forthcoming.

19. Ron Westrum, "Social Intelligence about Anomalies: The Case of UFOs," *Social Studies of Science*, (no. 7, 1977): 271–302; Westrum, "Knowledge about Sea-Serpents," pp. 293–314.

20. See note 18 above.

21. "The Loch Ness Monster," *King's College Hospital Gazette*, (no. 38, Spring 1950); vol. 29, no. 1, pp. 6–18. No identification of author is given, but it is attributed to Constance Whyte by Witchell, *Loch Ness Story*, 1974.

22. The stance of British zoologists is illustrated in Gordon B. Corbet, "The Loch Ness Monster," *Nature*, 15 Jan. 1976, p. 75; L. B. Halstead et al., "The Loch Ness Monster," ibid., pp. 75–76; Pearson Phillips, "Nessiteras Absurdum," *Observer*, 14 Dec. 1975, p. 11; J. G. Sheals et al., Statement from the British Museum of Natural History, as quoted in Robert H. Rines et al., "Search for the Loch Ness Monster," *Technology Review*, Mar./Apr. 1976, p. 37. For the cautious support of some Ameri-

can experts, see ibid., pp. 36, 37; Nigel Sitwell, "The Loch Ness Monster Evidence," *Wildlife,* Mar. 1976, p. 108. The anonymous survey is reported in J. Richard Greenwell and James E. King, "Scientists and Anomalous Phenomena," *Zetetic Scholar,* July 1980, pp. 17–29; James E. King and J. Richard Greenwell, "Attitudes of Biological Limnologists and Oceanographers Towards Supposed Unknown Animals in Loch Ness," *Cryptozoology,* (no. 2, 1982): 98–102.

23. See Tim Dinsdale, *Loch Ness Monster,* 4th ed., London: Routledge and Kegan Paul, 1982; Roy P. Mackal, *The Monsters of Loch Ness,* Chicago: Swallow, 1976; Nicholas Witchell, *The Loch Ness Story,* rev. ed., London: Corgi, 1982; Bauer, "Society and Scientific Anomalies," forthcoming.

5

The Evidence

The evidence proffered to support the existence of Nessies is anything but unambiguous. First, it is by no means obvious what actually is truly evidence and what it is evidence for: those who are convinced by chapter 1 will judge as relevant some things that seem irrelevant to those who are convinced by chapter 2, and vice versa. Second, no matter how chosen, the evidence is circumstantial. Third, it is not easy to get access to the original evidence. Even further, much of the evidence comes from personal testimony, which is not commonly granted much validity, especially not in science; errors, not to speak of frauds and hoaxes, have undoubtedly occurred at Loch Ness; and the whole business has many quite implausible aspects.

WHAT IS RELEVANT?

Inherent in the controversy is a lack of agreement about what evidence is relevant to the matter. For the prevailing conventional wisdom, it is relevant evidence that many people have mistaken ordinary objects for a monster; that hoaxes have been perpetrated; that the monster hunters are not teams of biologists; that the Loch Ness monster is frequently regarded as real by those who also believe in the reality of Bigfoot, UFOs of extraterrestrial origin, psychic phenomena, and so on. But for the monster hunter the only relevant evidence is the photographic, the sonar, and those eyewitness reports that have not been explicable as cases of mistaken identity or of deceit. That difference of viewpoints has far-reaching consequences and entails controversy and an associated hardening of opposing views.

By lumping together all that happens at Loch Ness, the conventional wisdom impedes the progress of understanding and the accumulation of certified knowledge about Nessies: any progress in knowledge requires and depends on analyzing, discriminating, classifying. Because of a preconception that the Loch Ness monsters might not be real animals but rather a conglomerate of hoax, myth, and mistake, the conventional wisdom does not practice the careful weighing of news reports and other data that is needed to separate wheat from chaff. Such discrimination thereby becomes ever more difficult as time passes. Through starting with the assumption that nothing is really there, the conventional wisdom also creates progressively more apparent justification for that assumption, as is evident in the following illustration.

In 1969 a photograph (fig. 10) was taken at Loch Ness of water movements that simulated a series of black humps. The photographer thought that Nessie had been captured on film and offered the photograph to a newspaper, which published it; the picture also

Figure 10. A photograph taken by Jessie Tait in 1969, in which one arm of a boat's wake simulates a succession of humps. Reproduced courtesy of the London Express News and Feature Services.

appeared on the cover of a guidebook that was widely available for a number of years.[1] Those uses of the picture likely confirmed for some unknown number of people that Nessies are figments of the imagination: actually, those "humps" are shadows on one arm of the V-wake from a boat; similar effects result when the wakes of two boats intersect, or when a single wake intersects the prevailing waves on the water. Now, the Loch Ness Investigation Bureau (LNI) had advised the newspaper not to publish the photo, pointing out that it showed ripples and not a Nessie.[2] Obviously, LNI saw publication as misleading the public and the newspaper's action as irresponsible journalism, more concerned with sensationalism than with accuracy. The newspaper, on the other hand, could well have seen LNI's advice as anything but disinterested, an attempt to suppress information relevant to one plausible interpretation of the Loch Ness phenomenon, namely, that it is largely or wholly a matter of mistakes by unsophisticated observers.

A CIRCUMSTANTIAL CASE

Even such a confirmed believer in Nessies as I now am has to admit that it is not yet, in 1986, totally unreasonable to remain unconvinced; after all, otherwise quite rational people are fervent disbelievers. (I would like to think that these statements will become badly dated during my lifetime.) There is no truly conclusive and direct evidence—the capture of a live specimen or the recovery of a carcass or a skeleton, after all, would settle the matter. A substantial reason for disbelief, then, lies in the fact that the best case for Nessie at this stage remains a circumstantial one. And circumstantial cases leave some room for doubt—when a murder has been alleged, even if someone confesses to it, one may, in absence of the corpus delicti, harbor doubts.

To find a circumstantially made case compelling one must be prepared to see coherence, a pattern of relationship, among phenomena that are not incontrovertibly related. Many judgments contribute to the final conclusion: judgments about which phenomena belong to the pattern and which do not; about the validity or plausibility of the facts or events individually; and about the plausibility of the pattern itself, compared to the plausibility of alternative patterns or to the possibility that no pattern at all exists. That such judgments are often made in part subconsciously and in part as a result of preconceptions makes it understandable that unproductive arguments are common between those who find a pattern and those who do not. All of us have had the bewildering experience of encountering individuals who see connections where we see none, and other individuals who will not see connections that are obvious to us. The difference in approaches between believers and disbelievers over the claimed existence of extrasensory perception provides an instance: the parapsychologists argue that where there is smoke there must be fire, to which John Wheeler has given the most direct answer—"Where there's smoke, there's smoke."[3]

When I first read Tim Dinsdale's book *Loch Ness Monster* (1961), I became intrigued because of the evident sincerity and veracity with which he recounted his filming of a large, moving hump in the loch; and I was anxious and ready for further evidence. But when the book turned to reports of aquatic monsters from other Scottish lochs and from Ireland and Iceland and Africa and Canada, I became incredulous: these were second- and third-hand accounts, from folklore and prehistory, much less convincing to me than Dinsdale's direct experience. I was not yet a believer, and so I was not ready to embrace a pattern that is in point of fact inevitable once one grants the reality of the Loch Ness species: for Loch Ness has existed in its present form for only some thousands of years, the animals must have colonized it from the sea, and similar lakes might well have been colonized too. Dinsdale, already a believer, was able to see the pattern that I could not. In later years, when I have confessed to others my belief in Nessies, I have seen the same reaction: some credence to my cumulation of evidence from Loch Ness but an initial unwillingness to add to that improbability the added apparent improbabilities from other places. Dinsdale has written of similar responses.[4]

I spoke earlier of the differences of opinion over which phenomena are relevant to the claimed existence of the Loch Ness monster. To go further here, I shall assume an agreement that known errors of observation and demonstrated hoaxes are left aside, and that the collection is agreed to include only what the monster hunters present as relevant: in other words, an agreement to leave aside the possibility that the actual pattern of coherence is one of mistakes by eyewitnesses, irresponsible journalism and tourist attracting, deliberate frauds, hoaxes, and the like.[5] In principle, one could test the relative probabilities that this is the true pattern, as against the existence of Nessies, by examining some of the data statistically. Of all reported sightings, how many have been shown to result from boats, or from their wakes, or from birds, logs, otters, etc.? Of all photographs taken of "the Loch Ness monster," how many are fraudulent and how many are of birds, logs, etc.? Such a procedure could in itself point to the likely solution. Thus many people judge as very low the probability that psychic phenomena are "real," because so many—though not all—of the claimed events have been proved to be trickery and deception. Or again, that 90 or 95 percent of UFO sightings have been explained as weather balloons, aircraft, planets, satellites, meteors, and so forth, convinces many people (including me) that there is nothing more than that to the whole business, that all the sightings could be so explained if enough information were available. Unfortunately the data do not exist for applying this procedure to the Loch Ness phenomena in a satisfactory way. It is the fact, however, that there are very few *demonstrated and published* instances of mistakes at Loch Ness by eyewitnesses who did not soon themselves recognize their own mistakes. No doubt they exist, but they have not been demonstrated in as wholesale a manner as in the case of UFOs.

The evidence to which I refer here, then, includes eyewitness reports; sonar results; photographic evidence; historical references, written or orally transmitted or as pictorial representations; and eyewitness and historical reports that Nessie-like creatures exist in other Scottish lochs, in lakes in other countries, and in the oceans. In assessing each type of evidence, judgments are continually called for. General problems which I discuss in following sections have to do with obtaining complete information; with the interpretation and reliability of eyewitness reports; with the jumbling together of reliable and unreliable statements in most of the literature; and with the varying judgments of different authors about the probability (or improbability) of particular claimed facts and of their possible relation to the Loch Ness phenomenon.

Here I want to emphasize only that there is already a judgment being made if one accepts that the "relevant" evidence of eyewitnesses, sonar, and photography all relates to the same entity (not to mention reports from history and from other localities). To the monster buff it is self-evident that the sonar and photographic results confirm the consensual parts of the eyewitness reports; the skeptic, on the other hand, points to the nonconsensual portions of the reported sightings and sees no need to believe that the sonar contacts have anything to do with what people have seen or photographed on the surface at Loch Ness. And there exist no general, accepted rules for deciding when two things are related and when they are not, before the connection has become a part of the conventional wisdom. "Shooting stars" or meteors were not connected with finds of meteorites before 1800; nowadays they are. Celestial patterns were connected with an individual's experiences in life at various times and places in the past; but nowadays few lend credence to astrological connections. Even the relation between sexual congress and the subsequent birth of a child remains unrecognized by some.

So I would suggest that it is not entirely perverse to view the monster hunter's collection of evidence as failing of incontrovertible interconnection. The sonar contacts prove no more than that objects that produce echoes exist, objects that echo as though they are quite large and move at (improbably) high speeds. The photographic evidence cannot readily be set aside—one cannot deny that unidentified objects, occasionally large, have in fact been photographed in Loch Ness. But there is a nontrivial step from that to interpreting the photography as all relating to the same thing,[6] and then inferring that the thing is a novel species of animal. (Even for the true believer, such a step is reminiscent of the attempt by a group of blindfolded individuals to ascertain the whole shape of an elephant by each touching only a small part of the creature; and even more so here because each photographic "touch" of Nessie may be of a different specimen, differing from others in age and size and gender.)

Not to belabor the point, a substantial reason for the lack of agreement about the existence of Nessies lies in the fact that the evidence is circumstantial rather than direct. One who adopts a posture of belief can continually and increasingly justify that belief by adding detail and context selected from reported sightings, photographs, folklore; but to the determined skeptic those additions are merely the conjoining of separate improbabilities, adding weak links to what was not a chain in the first place. A good illustration is the reported sensitivity of Nessies to sound, which has become something of a shibboleth for monster hunters. Among the fifty-odd eyewitnesses interviewed by Rupert Gould in 1933, one said that Nessie was apparently startled by noise, another that the creature submerged when an automobile sounded its horn.[7] During the succeeding decades suggestions of a sensitivity to noise have been no more frequently made by observers, again in only a few percent of the reported sightings. The skeptic is unlikely to be overwhelmed by that amount of evidence, but for the monster buff it enhances the plausibility that the phenomenon is animate; explains why sonar of low (that is, audible) frequency fails to detect large, moving targets; and indicates a need for silence while hunting (drift, do not use a motor; use blimps or autogyros for aerial reconnaissance, not helicopters or airplanes; do not slam car doors). The story of St. Columba from the sixth century is also adduced:[8] the saint commanded the water monster in Ness not to attack a swimmer and the monster obliged—the saint's loud voice acted on Nessie's sensitivity to sound.

Another shibboleth is the purported relation between frequency of sightings of Nessie and the number of salmon running through Loch Ness: *if* Nessies are predators, and *if* they normally feed on char and eels well below the surface, and *if* surfacings are accidental when a Nessie chases prey near the surface, then *perhaps* sightings are more frequent when the run of salmon is large. This is quite a nice speculative hypothesis, but to believers it has become a satisfactory explanation for a low rate of sightings in particular years.[9] Nowhere have I seen this speculation tested against the data of 1933, when sightings were at their highest frequency but the salmon fishery was at a low ebb.[10]

Similar shibboleths can be found throughout the literature on fringe subjects. Certain items become generally believed on quite slender evidence by the protagonists, for whom such details provide richness, a rounding out of the whole subject, a filling in of gaps, a starting point for theory and speculation. In studies of extrasensory perception, an example is the acknowledged decline over time in the ability of gifted subjects to guess what the order is of the cards in a pack: believers see this as consonant with any human activity—one gets tired—whereas the skeptics comfortably explain this as a predictable statistical effect if "giftedness" is merely a misinterpretation of the workings of chance. In those studies, and in parapsychology generally, similar shibboleths are the adverse effect of the presence of skeptics during experiments and the inability of gifted people to control their talents consciously. In ufology we have electromagnetic effects associated with close sightings and the assumption that some higher development of technology will permit forms of motion that currently appear impossible in the light of what we know about inertial forces. For Velikovskians the concept of "collective amnesia" is a case in point,[11] a phenomenon that has never been established as real but that would be capable of providing richly coherent explanations of a multitude of events—if only the phenomenon exists.

Now these examples are not intended to suggest that Nessies are not sensitive to noise but simply to illustrate how an initial difference of postures—belief or disbelief—determines that accruing details make the opposing views increasingly polarized and incommensurable. The protagonists develop a whole structure with its own inner logic, complete with tacitly accepted assumptions and even a special vocabulary; an outsider who does not accept the whole pattern is likely to see these corollaries as even less well established or more improbable than the central anomalous claim, and may be less likely to remain open-minded than would be the case if that person were exposed only to the hardest core of evidence. So, for example, I suspect that the invoking of St. Columba and similar stories is likely to make belief in Nessies seem implausible to individuals who might be inclined to inquire further if they were first exposed only to the photographic and sonar results.

Because the evidence is circumstantial, then, different people find in it different patterns or no pattern at all. Those who discern the "Nessie" pattern can invoke many points of confirming detail—which detail appears to others, who reject the "Nessie" pattern, as quite irrelevant or, at the least, not convincing "facts." Initially incommensurable attitudes become further entrenched, and it becomes increasingly difficult for believers and disbelievers to communicate in a meaningful way.

OBTAINING THE EVIDENCE

As one who became intrigued in the early 1960s by the possible existence of Nessie and took a decade to reach conviction about it, I feel particularly well qualified to speak to the frustrations of those who try to find out about the subject (though several books published in the mid-1970s have improved matters somewhat). One is not well guided toward the best information by encyclopedias, by science writers, by popular magazines, by newspapers;[12] instead, one must search out what the monster hunters themselves have written and said and done, and one must distinguish the honest and reliable monster hunters from the fraudulent or gullible ones.

The trouble is that (inevitably) no one of the published books tells the whole story. Nor is it a straightforward matter even to assemble a reasonably complete list of those books, in particular to hear of a new book when one is published. Nor is one likely to find many of these books in a given library—and a given collection may well include a preponderance of the unreliable. Some materials, newsletters, for example, are distributed only to members of certain organizations, and even the existence of those organizations has not always been a matter of common knowledge;[13] those useful materials are very difficult, even impossible, to obtain nowadays. And much remains to all intents and purposes unpublished. If one has the good fortune to attend a talk by one of the hunters, one hears about, and sees films or slides of, important things that are not in the literature. But not everyone can arrange to meet the hunters or to hear them talk.

More than two dozen books have been published that deal exclusively or primarily with Loch Ness monsters.[14] The frequently unreliable nature of those books and other factors that tend to make them less than completely convincing are addressed later; here the question is, how fully do they present to the reader the most significant evidence then available?

Gould, in *The Loch Ness Monster and Others,* did give a complete account of the meager data available as of 1934. In 1957 Whyte did similarly well in *More than a Legend,* a tribute to laborious collection of material from newspapers and through personal interviews. Dinsdale's 1961 book, *Loch Ness Monster,* was, of course, the first to publish stills from his film; it also included two new photographs—the O'Connor and Cockrell photos (whose validity, however, is not universally acknowledged)—but failed to include the Gray or Stuart photos or the Irvine stills. (Later editions in 1972, 1976, and 1982 omitted one of the controversial photos and added the Lowrie wake.) Published a few months after Dinsdale's book was one by Maurice Burton, *The Elusive Monster,* of which more will be said later (see chap. 7); for the present purpose it is sufficient to note that a reader of Burton's book would never suspect that Burton knew Dinsdale. In generally arguing against the reality of Nessies, Burton failed to publish the Macnab photo and gave an entirely misleading sketch of the Gray photo. Holiday, in *The Great Orm of Loch Ness* (1970), failed to give the Stuart or Macnab photos or stills from Dinsdale or Irvine; Baumann's *The Loch Ness Monster* (1972) has no stills from Dinsdale or Irvine, nor the Lowrie wake, and shows the Gray photo upside down. The otherwise very good book by Campbell and Solomon, *The Search for Morag* (1972), has no Dinsdale or Irvine stills, nor the Gray photo, nor the Lowrie wake.

Quite in general, then, the reader of books about Loch Ness is not necessarily shown the best photographic evidence available; and some recent efforts are even worse in this regard, particularly the potboilers by Akins (*The Loch Ness Monster* [1977]), Landsburg (*In Search of Myths and Monsters* [1977]), and Cornell (*The Monster of Loch Ness* [1977]). Costello has good coverage in the bibliography and tables of photographs of *In Search of Lake Monsters* (1974), but he failed to reproduce the Lowrie wake or the very important "flipper" photo. Mackal's *The Monsters of Loch Ness* (1976) has very good coverage of the literature, a table of more than 250 reliable sightings, and discussion of many photographs and moving films, but no Dinsdale or Irvine stills are actually reproduced. Witchell's book, *The Loch Ness Story* (1974), particularly the 1976 second edition, is hard to fault, however, in completeness of photographic evidence. (My own experience in attempting to secure permissions to reproduce photographs for this book inclines me now to be more charitable toward those who preceded me in the endeavor.) So, at last, one can find satisfactorily complete presentation of the evidence, photographic in particular, through referring to Mackal and Witchell; the difficulty now for the unsophisticated reader is to know that to be the case, since libraries are just as likely to have less satisfactorily complete books than these on their shelves. From about 1960 to the mid-1970s no single book gave anything like comprehensive coverage.

The searcher's frustration nowadays is likely to focus on material that is said to be available and yet has never been made public. The failure to publish relevant material and the disappearance of important evidence as time passes are inevitable corollaries of the fact that Nessiedom is not yet science, that there is no such repository for material about Nessies as exists for items in the mainstream of science, with its widely available journals and comprehensive sets of abstracts and indexes. So the moving films obtained by Irvine in 1933 and 1936 were lost, leaving us with only a couple of stills whose credibility or value is much less than if the whole films were extant.[15] So too with the missing Fraser film of 1934;[16] and the negative of the Gray picture is lost so that computer enhancement is not possible.[17]

Those who enjoy digging in libraries will also be frustrated at the failure, in much of the Nessie literature, to cite references in a form that makes it possible for the references to be located again. Thus a reporter came across an old Scottish lay dealing with the monster of Loch Morar but failed to note where.[18] John Keel mentioned a full-page article about Nessie in the *Atlanta Constitution* of November 1896; others, including me, have scanned every issue of that month without finding it.[19] Someone claimed a reference to the "leviathans" of Loch Ness in Daniel Defoe's *Journey Through the Whole Island of Great Britain* (7th ed., vol. 4, 1769), but no one else has been able to confirm that.[20] Costello, himself frustrated at being unable to confirm the *Atlanta Constitution* and Defoe references, tantalizes his readers with a reference to the *Glasgow Evening News* of 1896[21]—could he not have given a more complete date, if not the page? I have read a number of times that Chesterton once said that men have been hanged on less evidence than there is for the Loch Ness monster;[22] but none of the reports let me know where I could verify that quote or determine its context.

Daily newspapers published the now-classic "surgeon's" photograph in 1934. Surely it became known in their negotiations that the surgeon had taken a second successful picture at that time! Yet the public came to know of that only twenty years later, when Constance Whyte found the man who had developed the original plates and had kept a contact print of the second shot.[23] The grain on that print makes it impossible now to enlarge it to the same extent as the classic one was, from the plate itself; so possibly decisive evidence—at least about whether it is a Nessie, a bird, or an otter's tail—is lost to us. The monster buff could read, in the early 1970s, that a computer enhancement of the first photo shows "whiskers" at the jaw; but that computer enhancement has not been published, though it continues to be referred to.[24]

The Lowrie wake was photographed in 1960 and has been included in books by at least Dinsdale, Mackal, and Witchell.[25] I had tried to follow news about Nessie since the early 1960s and joined the LNI when I heard of it in the mid-1960s. Imagine my surprise when, in the mid-1970s, I heard a talk by Sir Peter Scott (at a Wildlife Federation meeting at the Galt House, Louisville, Kentucky, 20 Mar. 1976) and not only learned for the first time that more than one photograph had been taken but saw slides of two more of those Lowrie photographs.

The "flipper" photograph of 1972 was widely taken by monster hunters to be a major advance, perhaps the achieving of scientific respectability for Nessie; even *Time* ("Myth or Monster?" 20 Nov. 1972) published it. But the public was not then told that a *second* "flipper" photograph had been obtained, despite the fact that it adds so much plausibility to the interpretation that the "flipper" is indeed an appendage of a moving animal. Dinsdale's article of 1973 makes no mention of the second photograph; and the Ness Information Service newsletter expressed surprise, in 1976, that this second "flipper" photo even existed.[26] There has also been reference to another photograph, obtained during a series of experiments in 1972, that supposedly shows a tail-like structure; to my knowledge it has never been published.[27]

The underwater photographs of 1975 are tantalizingly suggestive yet not quite clear enough; one cannot help wondering how much they might be improved by computer enhancement. One reads that this has in fact been tried but does not find the results published.[28] That the Academy of Applied Science obtained at least one intriguing underwater photo in 1981 has been communicated only in a 1982 report to its membership.

The systematic surface watches organized by LNI, with very powerful photographic apparatus at the ready, failed to produce the definitive proof of Nessies that had been the original aim. Short pieces of film were obtained, however, and were mentioned in bulletins to LNI members and occasionally in newspapers and magazines and books, but none of the films was made publicly available even in the form of stills.[29] The LNI did not publish these fragments for the stated reason that they were not in themselves conclusive; but mentioning material that is not publicly available not only carries no conviction in scientific circles, it even arouses suspicion. The fact that Mackal mentions other scraps of LNI film that were later found to show birds rather than Nessies detracts further from the overall credibility.[30] The scientifically inclined would want to see *all* these films, to judge for themselves how plausibly some but not others can be said to show birds.

Failure to publish in readily accessible, standard periodicals continues. The Smiths took an 8mm movie in August 1977 of a pole-like object that came out of the water three times, submerging intermittently; the film was eventually shown on television in England, in February 1979, but no stills have yet been published. The *Nessletter* of August 1977 mentions photographs taken by a certain McGrew, which have yet to appear in print; it also mentions a photograph published in *Fortean Times,* not a readily available periodical since few libraries hold it. On several occasions we have been tantalized by reading that hydrophones have detected clicks, chirps, and knocks, coming from no known fish or eels, but we wait in vain for full publication: duration of sounds, modes of analysis used, available knowledge about underwater sounds made by known species—all the data needed for judging the possible significance of the reports.[31]

Altogether, then, there has been incomplete and fragmented publication of the data from Loch Ness, making them difficult or impossible to obtain, let alone to assess. In one sense this is a criticism of the monster hunters, that they have not made the best public case with what is available to them. But that is too superficial a view. The generally valid truth comes down to the fact that Nessies are outside the professional and conventional wisdoms, and that in itself severely restricts the opportunities for publication of the monster hunters' data. Such plausible outlets as *Nature* have rejected important articles, for example, the sonar results subsequently published in the *New Scientist,* a respectable enough magazine but with less standing in the scientific community than *Nature.*[32] And *Science* as well as *Nature* rejected studies of the eel population in Loch Ness, in one case despite what Mackal informed me were favorable referees' reports.

What has long been needed and is still lacking is the publication of results as they are obtained, with strict editorial control to ensure that all relevant details are described, loopholes in reasoning plugged, gaps in data filled in or at least acknowledged—in other words, the procedure of publication that is characteristic of science, which ensures that the literature on a given subject meets certain standards and represents, over the course of time, a progressive cumulation of material of increasing reliability. But the scientific literature is not open to Nessie, so data have the appearance of being withheld and what is actually published does not meet standards that are normal in science.

Rines's team published quite a detailed account of their results from 1972 and 1975 in *Technology Review,* thus making available important photographs with the imprimatur of a respectable periodical.[33] But theirs is a popular account, not a scientific paper. The typical scientist refereeing the article would have demanded more information on several issues: calculation of the object-to-camera distance in detail, with estimated limits of error; programs used in the computer enhancement; the number of frames of film that were entirely blank, the number that showed eels, fish, or other common objects, and the number (if any) of those that showed unexplained objects but were not reproduced in the article; and so on.[34] I am not suggesting that all this could in fact have been done; indeed, it would have turned the article into a monograph. But had the scientific literature been open to the subject in the past, much of the needed material would have been published earlier—all the results from 1972, for example—and the new results could have been treated at manageable length in impeccable detail.[35]

It is simply the obvious vicious circle: until Nessie becomes science, the literature on Nessie will not meet the standards of science; and that the literature does not meet those standards impedes the acceptance of Nessie into science. The extent to which that is true is illustrated by the aborted symposium of scientists scheduled for 1975 in Edinburgh: a few unauthorized mentions in the press of new photographs to be shown were enough to cause the scientific establishment to cancel the symposium on the grounds of prejudicial publicity. The experienced monster hunters have learned through bitter experience that biologists will look at evidence for Nessie only if they can do so in complete, unpublicized privacy. When such conditions are scrupulously observed, however, as by Dinsdale in 1960 for months,[36] all that happens is that the experts say, "Interesting, but inconclusive; go and do better." Yet the doing better is impeded by lack of funds and organization, which might be obtainable if the subject had the imprimatur of science, which science refuses publicly to give. So the monster hunters are caught again: if they publish in the media open to them, the biologists are less likely to look seriously at the data; if they do not publish, they lose the potential support of the community of monster buffs and of the independent or eclectic patron.

EYEWITNESS TESTIMONY

One powerful reason for disbelief in Nessie is that so much of the case has been built on eyewitness reports: the fallibility of human observation is an entrenched article of faith in the conventional wisdom. And there are excellent grounds for subscribing to that article: for example, the classic demonstration in which someone rushes into the classroom, shoots the teacher, and runs out; the subsequent descriptions of the intruder by members of the class turn out to vary across the available spectra of short to tall, blond to dark, clean-shaven to bearded.

Surely we all have our own experiences of making incorrect interpretations at times. About 1970 three friends and I saw a UFO over Lexington, Kentucky. It was late afternoon, under cloudy gray skies, and we all agreed that we saw a disk with lights around the circumference, flashing on and off in sequence. We chased the thing by car for five minutes without changing our interpretation. Later that evening it returned, passing directly overhead, whereupon we all realized that the lights spelled out "Goodyear"—it was the blimp, cruising on the eve of a football game. In 1958 I saw a sea serpent in the Indian Ocean: a succession of black humps, many miles from the ship, and across the whole horizon no trace of another vessel, so this was no effect produced by waves and wakes. But through 20X binoculars there *was* another ship to be seen, miles from us and miles from the "sea serpent"; the latter was clearly, at that magnification, the intersection of wake and waves. On another occasion, fishing in the ocean, standing on rocks at sea level, I also saw a sea serpent: a large black neck forming half of a coil, thickening at one end where part of the body showed; black, smooth, shiny. Yet a few minutes later the head appeared and the bird flew away—I had been entirely wrong about distance and size.

At Loch Ness the opportunities to be deceived are legion. In flat calm, gusts of wind can give startlingly realistic approximations of sharply localized disturbances, of apparent wakes caused by nothing visible. When the water is relatively smooth and the wind gusty but not too strong, the wind-ruffled surface has a silvery appearance, and calm patches stand out as black "humps" of various sizes and shapes, changing sometimes slowly and sometimes fast, sometimes moving, often also giving the appearance of water being disturbed at the edges of the "hump." When there are small wavelets on a generally calm loch, their tops frequently look like rounded or angular black bodies knifing through the water, coming up and diving again. The water birds also perform, drifting or paddling along, sometimes with only head and neck showing, diving periodically to reappear much later, farther away. These birds can produce impressive wakes, too, particularly when they take off in flight; seen from behind, the bird is readily lost to sight as it leaves the water—so we have a mysterious wake, caused by a black something that then disappears (dives!). . . .

The wakes left by boats also produce quite tantalizingly mysterious effects as they run across the prevailing waves. These boat wakes persist for an incredibly long time—one must see this to believe it—as the line of waves moves sideways across the mile of loch surface, long after the boat is out of sight (it can take twenty minutes for waves made by a boat in mid-loch to reach the shore).[37] The hopeful observer must be armed with a good pair of binoculars to avoid being deceived by these and other phenomena at ranges of a few hundred yards to a mile or more. And looking up or down the loch, which is more than twenty miles long, deception at long range cannot be guarded against. Several instances of deceptive appearances have already been mentioned in chapter 1.

On both general and particular grounds, then, skepticism about the reliability of eyewitness reports is well founded. Since most of the detailed case for Nessie has been presented on the basis of human observation, it is not surprising that a general air of doubt clings to the subject. The monster hunters have been well aware of the problem. Thus Gould took pains to discuss the question, emphasized the care he took in questioning the witnesses,[38] and gave their full names. Whyte knew her eyewitnesses personally and assured the reader of their trustworthiness. It is in the nature of things, however, that such assurances carry weight only with the few readers who personally know the author, while they give free play to the skepticism of the cynics who do not. So, when Whyte's witnesses are identified by initials only, not by full name, the skeptics take their case as nearly proven. Yet it was kind, indeed necessary, for Whyte to guard her informants against the merciless ridicule and badgering to which sighters of Nessies and sea serpents are subjected.[39] In 1957, when her book was published, some libraries at first refused to buy it. Bishop Hoffmeyer, visiting in Edinburgh, found the bookseller smiling when he sold him Whyte's book, a review of which characterized the prevailing attitude thus: "An aura of imposture, the echo of innumerable music-hall jokes, is a main ingredient in the aspic wherein our mental image of the Monster is preserved."[40]

Dinsdale found reassuring the consistency of the eyewitness reports (except on the number of humps). But skeptics reply that ever since 1933 people have known how the "monster" is supposed to look and naturally "see" much the same thing. General skepticism about visual observations has doubtless been reinforced by the inaccuracy and sensationalism with which observers' accounts are reported in the press. Several authors have pointed to the restraint and plausibility of the statements made to them by observers of Nessies, in contrast to the manner in which those same observations were reported by the newspapers.[41]

At any rate, anecdotal accounts by lay human observers have little evidential power in science, and the conventional wisdom follows that lead. Much of the case made for Nessie has suffered, I believe, from the fact that primacy has been given to the eyewitness reports, adducing photography and sonar as confirmation, instead of taking the opposite tack—emphasizing the uncontroverted and unexplained movies, still photographs, and sonar contacts, and using the reported sightings only for added detail. The rationale for the believers' approach is clear, of course: from the accumulated hundreds of well-attested sightings, a much more coherent and detailed description can be built of what the animals look like and of how they behave, than from the fragments of photography. But it is a fact of life that such a case will never be acceptable to science. The ambition to persuade scientists that they should be prepared to accept the testimony of human witnesses as scientific evidence seems to me to be whistling in the wind.[42] It is not here a matter of possible dishonesty but of the manifold opportunities for people to make honest mistakes of memory and in perception. If a phenomenon is reproducible, or fits into the scheme of what is already known, the scientific community readily accepts also the statements of eyewitnesses, of course. But it is entirely different to seek to establish primarily through human testimony the scientific validity of something improbably anomalous.

I say this even though the Loch Ness story is a striking demonstration of the reliability that can be achieved when common sense and judgment are used, by an intelligent person, to assess eyewitness reports. Consider the picture of Nessie assembled by Rupert Gould in 1933 from his interviews of about fifty individuals. Subsequent work has lent nothing but confirmation to that description, but this really tells us only that, in this instance, Gould's weighing of eyewitness testimony was sound. There is no way to make routine and objective the reaching of such judgments; that Gould's judgment was sound here does not mean that it would have been sound in other situations, nor that the whole community of scientists—even assisted by lawyers—would be capable of exercising sound judgment about eyewitness testimony in any given case.

THE WHEAT AND THE CHAFF

Not only is it difficult to assemble or gain access to all the hard evidence claimed by the monster hunters, but there are also disagreements among them about the validity of significant items.[43] Those disagreements, inevitable as they are, make the whole case less convincing to the outsider. Even worse, some of the books and articles are unreliable on plain matters of fact. And there is no mechanism for controlling the quality of the literature, no publicly established and widely known authority to ask which books are reliable and which are not. The intrigued inquirer has only personal judgment to fall back on, judgment that is bound to be uninformed until he has read and seen a great deal—the difficulty of doing which I have just discussed. So we have a vicious circle, inevitably present when a subject is not in the mainstream of science. There, writers are under an obligation to have mastered the earlier writings of others; and reviewers check manuscripts for competence and reliability before they are published. No such mechanisms exist in Nessiedom (or in other fringe fields) to safeguard the quality of the literature and to make that literature progressively more informative.

The Gray photo, accepted by most believers as validly of a Nessie, suffers from having been first published in severely retouched form; from being published upside down in some cases; and from a drastic misrepresentation by Burton in a purported sketch of the picture. One of the Irvine films has been obliquely characterized as possibly fraudulent. The surgeon's photograph, accepted almost universally by believers as showing a Nessie, is said by Mackal to show a bird, while Burton has vacillated between regarding it as genuinely of an unknown species of large animal and characterizing it as the tail of a diving otter; allegations that it is a hoax remain inconclusive. Again, Mackal has cast doubt on the Macnab photograph, which is regarded as genuine by others.[44] Not unreasonably, then, the skeptic and the cynic can call into question a significant portion of the best available data, and the cited disagreements readily support a totally disbelieving stance. If Mackal the scientist finds ground to question two of the four or five "best" surface photographs, perhaps he himself was a bit gullible about the others, and *all* the evidence can be dismissed. . . .

The existence of many other purported photographs of Nessie adds further confusion. Leaving aside the certain frauds for the moment, there remain a few more pictures that are claimed as evidence by some monster hunters and rejected by others. The

O'Connor photo (fig. 11) was published by Dinsdale in *Loch Ness Monster* (1961) but omitted from later editions;[45] Costel'o and Rabinovich lend it credence, Mackal rejects it, Burton claims it a hoax constructed from plastic bags that he found at the loch. The Cockrell photo (fig. 12) is generally described as a "maybe," though Mackal unequivocally calls it a log. A photograph (fig. 13) published by Burton, Costello, Mackal, and Witchell shows nothing to indicate that it was taken at Loch Ness; it is grainy to an extent that indicates very great magnification, and there is even argument over who the photographer was. Fraser's film of 1934 was stated by Burton to have been lost, with not even a single frame preserved; yet a still (fig. 14) claimed to be from that film was published in 1970 and a similar sketch was also published.[46]

Discrepancies and disagreements on other matters than the photographic evidence also abound. In 1980 I counted twenty-two books in English dealing directly with Loch Ness monsters, eight of them specifically written for young readers. I could recommend no more than half of those twenty-two as generally reliable.[47] Gould's *The Loch Ness Monster and Others* (1934) and Whyte's *More than a Legend* (1937) are reliable but dated. Dinsdale's *Loch Ness Monster* (1961, 1972, 1976, 1982), *The Leviathans* (1966), *Monster Hunt* (1972), *The Story of the Loch Ness Monster* (1973), and *Project Water Horse* (1975); Witchell's *The Loch Ness Story* (1974, 1975, 1976, 1982); and Mackal's *The Monsters of Loch Ness* (1976) are sound, though Mackal takes an idiosyncratic stand on several of the classic photographs and makes a few more or less serious errors: that the "flipper" pictures were taken with sonar-triggered cameras which, in point of fact, have yet to give usable results (p. 111); that the Academy of Applied Science is affiliated with the Massachusetts Institute of Technology (p. 111). Another book, Searle's *Nessie* (1976), is by a man whose integrity has been brought into real question, and Burton's *The Elusive Monster* (1961) is an exemplar of specious reasoning, as is the recent book by Binns, *The Loch Ness Mystery Solved* (1983).[48] Holiday dabbles in the occult connection in *The Great Orm of Loch Ness* (1970) and *The Dragon and the Disc* (1973); and both Akins's *The Loch Ness Monster* (1977) and Landsburg's *In Search of Myths and Monsters* (1977) are merely inferior potboilers with a sprinkling of errors.

Figure 11. A sketch of the controversial photograph taken in 1960 by Peter O'Connor. Efforts to obtain permission to reproduce the photograph itself were unsuccessful. It has been published in several places, including: Costello, *In Search of Lake Monsters* (1974); Dinsdale, *Loch Ness Monster* (1961); Mackal, *The Monsters of Loch Ness* (1976); and Rabinovich, *The Loch Ness Monster* (1979). Sketch by Ron Garrett.

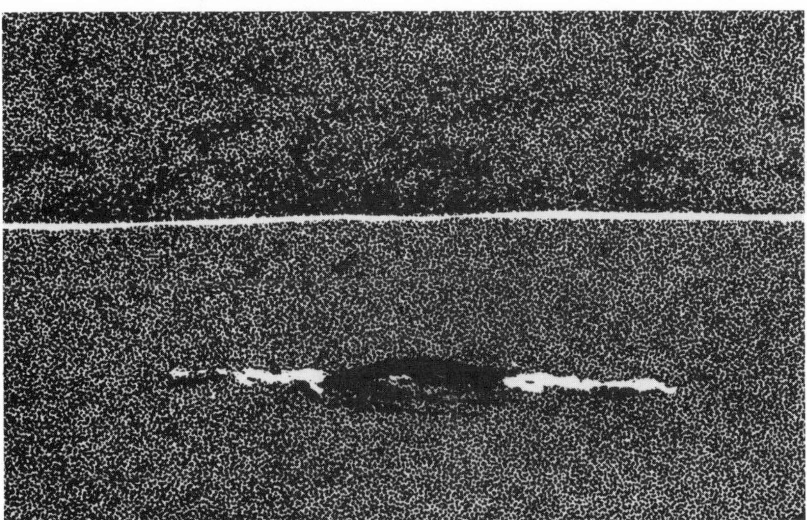

Figure 12. Is it a Nessie or a floating stick? A sketch of the Cockrell photograph of 1958. Mr. Cockrell gave permission to reproduce the photograph itself, but the agency handling it demanded what this author regarded as an exorbitant fee. Sketch by Ron Garrett.

Figure 13. A photograph published in the *Illustrated London News* on 1 September 1934; circumstances unknown, photographer not known with certainty. Reproduced courtesy of *Illustrated London News.*

Figure 14. A single frame published in 1970 and supposedly from the film taken by Captain Fraser in 1934; provenance doubtful in the extreme. Reproduced courtesy of The Photo Source.

Most dangerous, perhaps, is the book by Costello, *In Search of Lake Monsters* (1974). On the one hand it has good coverage of the photographs and good references and bibliography; on the other hand the text is replete with errors ranging from mere sloppiness to absolutely wrong statements. We are told that the story of St. Columba at Loch Ness in A.D. 565 is given in an eighth-century (p. 25)—or ninth-century? (p. 15)—biography written "a century after his death" (p. 25). Heuvelmans appears as Heuvlemans (p. 109), Rines as both Rine and Rhine (p. 111), and the reader can also choose among Koldewey, Koldwey, and Koldeway (pp. 245, 246). We learn that from June 19 to June 27 is five days (p. 88), and that Loch Ness is a mysterious heat trap—though Costello found the water cold (p. 23), it is actually forty-two degrees *centigrade!* He tells us that the second surgeon's photo shows a smaller object than the first (p. 53), when in point of fact the difference is in the degree of enlargement from the original plates;[49] that the body and neck in the controversial O'Connor photo resemble those in the questionable Adams/Lee photo (p. 88; see fig. 13), which shows no body at all and doubtfully a neck; that Tiamat on a Babylonian cylinder seal shows the same animal as in the surgeon's photo (p. 252); that the remarkable Swordlands sightings at Morar were of the "upturned boat" type (p. 149), when actually they were of a three-segmented underwater shape;[50] that a mane and fur are often reported on Nessies (p. 284); that the antennae reported three times (p. 118) are actually ears, sixteen inches high by eight inches long (p. 203).

About half of these books, then, do gross disservice to the reader and thus also to the publicly made case for the Loch Ness creatures. Children just beginning to read have the choice of two books, one of which (Bendick's *The Mystery of the Loch Ness Monster* [1976]) is thoroughly reliable, the other (Rabinovich's *The Loch Ness Monster* [1979]) nothing short of a disgrace—the Gray and the Rines "gargoyle head" photos printed upside down, the latter and the "body-neck" photo inverted left to right, the controversial O'Connor photo cropped to remove the neck entirely; a statement that the surgeon's photo shows a huge body; a ludicrously wrong illustration of how the land rose to cut Loch Ness off from the sea; and "Nessiteras" is spelled "Nessitara" . . . all in just forty-eight pages.

For older children there are five books to choose from and only two good choices to be made—Cooke and Cooke's *The Great Monster Hunt* (1969) and Dinsdale's *The Story of the Loch Ness Monster* (1973). Baumann's *The Loch Ness Monster* (1972) has the Gray photo upside down, ascribes to Loch Ness some historical references that actually refer to other lochs (pp. 3, 45), and manages to combine jocular insult with considerable inaccuracy when stating that "after many years of scoffing at the idea of a monster in Loch Ness, the scientists finally began to scratch their beards and give the matter some serious thought" (p. 43). Cornell's *The Monster of Loch Ness* (1977) gives no references and is quite misleading: there is a photograph by a fraud (see chap. 6) who had been exposed several years before the book appeared; the alleged reference to "whales" seen at the loch in the eighteenth century is given bluntly as fact and without citation of source (p. 7); and a sonar image of an unidentified structure on the bottom of the loch is said to be of a monster's carcass. Snyder's *Is There a Loch Ness Monster?* (1977) is poorly organized: for example, it treats the O'Connor photo as possibly reliable in one place (p. 83) but as a fraud in another (p. 161), where the date is also incorrectly ascribed.

This mishmash of sound and unsound, responsibly cautious and irresponsibly sensational, it should be recalled, represents the literature to which the inquiring mind would most naturally turn, the books directly concerned with Loch Ness. Articles in magazines, newspaper items, and brief mentions in various places present readers with a similar concatenation. The very state of the published material about Loch Ness, therefore, is an excellent reason for people not to take seriously the possible existence of Nessies.

It is worth noting also that the hundreds of thousands of tourists who have passed by Loch Ness have been exposed to no more reliable a selection of information. Already in 1933 fake pictures of Nessie on postcards were taken by many people to be of the genuine article. During the late 1960s and early 1970s the LNI disseminated relatively sound information during the summer from its camp at the loch-side; but in 1980, for example, visitors had a choice of three exhibits dealing with the monster—two of them (at the Drumnadrochit Hotel and in the Great Glen Exhibition at Fort

Augustus) quite sound but the third quite fraudulent (see chap. 6). The Inverness Public Library is quite small and does not have good coverage of the literature about Nessies; what it does have (as of summer 1980) combines the unsound (e.g., Costello's *In Search of Lake Monsters;* Searle's *Nessie*) with the reliable (e.g., Dinsdale's *Loch Ness Monster,* 3d ed.; Whyte's *More than a Legend;* Witchell's *The Loch Ness Story,* 2d ed.). And the guidebooks on sale everywhere also cover the gamut from good to bad.[51]

In 1973 I purchased examples of the only three pamphlets I could find. The cheapest—in price and quality of presentation—was by Father Carruth, published by the Abbey at Fort Augustus. It is accurate enough as far as it goes but lacks references to some of the most important evidence and major books, and it contains no photographs. Then there is the booklet published by J. Arthur Dixon, profusely illustrated in color on glossy paper. The text is quite good, though again lacking references; but together with a couple of the important classic photographs, there is on the cover and as a centerfold a photograph (see fig. 10 in chap. 5) of an unmistakable *boat* wake, labeled "Is it a ripple, or . . . ?"—a caption likely to promote skepticism even in one desirous of believing. Finally there is the booklet by Barrie Robertson, with black-and-white illustrations, well presented on good glossy paper, featuring several of the photographs promulgated by Frank Searle. Some of those photographs are undoubtedly spurious, and none of them can be accepted as genuine (see chap. 6).

By 1980 the general reliability of guidebooks had not improved. A new edition (1975; rev., 1976) by J. Arthur Dixon is good and reliable. Also quite good is a booklet by William Owen, and the new edition by Father Carruth mentions more of the important photographic evidence than did earlier editions. But the *Handbook—Loch Ness Monster* by Jim M. MacRae (Inverness, 1979) describes as a "dedicated full time monster watcher" the aforementioned Frank Searle and reproduces one of his doubtful photographs (characterized as "one of the best ever taken"). Similarly misleading is *Loch Ness Country by Car* (1977), one of a series of very detailed guides to touring, with accurate strip maps and mention of places worth visiting. In it Searle is given credence for 21,000 hours of watching, twenty-six sightings, and some eight photographs; and his (later successful) battle with the local authorities is mentioned, for permission to locate his Information Center at the loch-side.

In 1985 the mixture was much the same. The Loch Ness Monster Exhibition at the Drumnadrochit Hotel had been greatly expanded and improved; the curator plans to establish an archive accessible to serious students, a considerable advance in making reliable information publicly available. On the other side, there was a new exhibition, the Loch Ness Monster Research Exhibition, which had opened in 1984 in Inverness. It showed little more than the underwater photographs of 1972 and 1975, but more than made up for the lack of data by far-reaching interpretation—claiming, for instance, to identify the particular *species* of plesiosaur that Nessies might be, and that the sonar results prove the existence of two twenty-eight-foot creatures and one nineteen-footer. It was the rankest speculation, presented to all as fact and clothed in technical jargon whose authoritative tone no doubt seduced some of the unwary. More happily, in 1985 there was no sign of Frank Searle at the loch, though some of his photos were on sale at the Falls of Foyers.

Little new literature was locally available in 1985: no new pamphlets and only very minor revisions of the old ones. It is largely from these cheap publications, and from the exhibitions, that visitors get their information. The exhibition in Drumnadrochit logged 120,000 visitors in 1984, making it one of the top twenty attractions in Scotland;[52] by contrast, very few inquiries are made about Nessie at the Inverness Public Library or at the Inverness Museum.

LARGE AND SMALL IMPROBABILITIES

Faith may be defined briefly as an illogical belief in the occurrence of the improbable.

H. L. Mencken

The very existence of Nessies is improbable enough, in the sense that their existence would not be inferred on the basis of contemporary knowledge in biology. But some of the evidence adduced to support the case for Nessies makes the whole business seem even more improbable rather than less. Certain of these extreme improbabilities stem from individual judgments by various writers, but some of them seem to be inherent in the nature of Nessies.

A considerable but unavoidable embarrassment to the most hardheaded hunters is the existence of a small number of reports of Nessies having been seen on land. In Nessiedom these events have a place that is not unlike that of the "close encounters of the third kind" in ufology. One is brought squarely up against the phenomenon of apparently responsible and plausible individuals who insist on the reality of experiences of the most extremely improbable sort. The number of claimed Nessie sightings on land is between one and two dozen, depending on how one judges the reliability of the individual accounts. Unfortunately there is no ready way of dismissing them all as misidentifications of, say, otters. Indeed Burton, who is prepared to discount virtually all the evidence that most others regard as persuasive, finds himself left with the need to countenance the possible existence of a twenty-foot-long otter-like species that would best be looked for on land rather than in the water.[53]

The credibility of the Loch Ness monster was bound to be damaged by the inherent improbability that an unknown species of large aquatic animal, seen infrequently enough in the water, would disport itself at times on land. Moreover, this embarrassing improbability entered the story at the very beginning. Already in July 1933 a sighting on land was reported, with quite indefinite and confusing descriptions of what had been seen; and a second land sighting was reported in January 1934, with a quite different description of the creature involved as well as questionable doings about traces left on the ground on that occasion.[54] So appearances on land have been an integral part of the affair from the start, and belief in the existence of Nessies is surely made less easy in consequence.

Another whole set of improbabilities is introduced by references in old records and legends. I have already pointed out that the story of St. Columba has appreciable evidential value for believers but perhaps the opposite effect on others. In our day the tendency is to look with distrust on the opinions of past generations and to assume that legends have their roots primarily in the imagination rather than in reality. So, I would suggest, the public view of the case for Nessie has suffered rather than benefitted from the purported early accounts, particularly when the citations turn out to be inaccurate or unconfirmable and when they sound less like factual reports than like stories of water-horses, water-bulls, and water-kelpies with magical attributes.[55]

In the same vein, I believe that acceptance of the case for Nessie has been hindered by the conjoining to it of reports of more or less similar creatures in other parts of the world. A postulated relationship of Nessies to sea serpents surely hurts Nessies in the eyes of the public; with less direct evidence to establish the point, I would suggest that similar harm was done by bringing in tales from Iceland, Sweden, Canada, Siberia, and so on.[56] The evidence from all those other places, after all, is more tenuous than that obtained at Loch Ness. Moreover, the fact that stories of such animals are associated with many other Scottish lochs[57] and with many Irish loughs means, to the skeptic, that these are all, including Loch Ness, mere carryovers of Celtic myths. And even the described appearance and behavior of at least some of those other animals really differ quite drastically from those of Nessies, so that one feels one is being asked to believe in the existence of not just one new species but several. The tales from Ireland in particular stretch the hardiest believer's credulity to the utmost. Nessie's reputation cannot have been enhanced when some of the hunters sought similar creatures in Irish waters barely large enough to make a bathtub for one Nessie[58]—even though the search was not entirely fantastic, given the hypothesis that Nessies are marine creatures that occasionally visit these readily accessible loughs.

At any rate, there are several major improbabilities inherent in the whole business. It is thereby more difficult to have the case listened to, let alone accepted, than if there were only the one fundamental improbability of the Nessies themselves. Moreover, some gratuitous improbabilities have been added to these inherent ones. Holiday stretches the imagination when he concludes in *The Great Orm of Loch Ness* (1970) that Nessies are invertebrates, giant types (twenty or more times larger) of Tullimonstrum, a fossil conveniently discovered not long before he wrote his book—albeit the fossils had been found in the center of the United States and would not appear to most people to have even a resemblance of shape to the Nessies. Holiday asks us also to believe that there have existed since the 1930s, in a bank vault and under guard by a secret trust, two films of Nessies, one taken at the loch and the other at a sea-loch, a story denied to others by the "guardian" Holiday quoted.[59]

Holiday's later book, *The Dragon and the Disc* (1973), is a concatenation of far-fetched ideas. He talks of the wisdom of the ancients as "still in advance of formal scientific attitudes." Nessies are organically related to UFOs (because, he claims, dragons were in the past), which themselves have some relation to megalithic structures, the ley lines joining which are correlated with the Martian canals.[60] He claims that mysterious power has enabled Nessies to remain undiscovered: they maneuvered around the range and field of view of the LNI cameras;[61] or, when they failed to do that, they caused the camera to malfunction. And if that was not successful, they arranged it so that the films would be locked away inaccessibly in a bank vault! I talk later (in chap. 6) about the unfortunate effect on Nessiedom of the coupling of that improbability with other fringe subjects; here the skeptics can point to at least one man who was a determined monster hunter and also gullible in the extreme about a number of occult matters.

Notes

1. The 1969 photograph by Jessie Tait appeared in Ron Lyon, "Could This Be the Loch Ness Monster?" *Sunday Express,* 7 Sept. 1969, p. 7; *Loch Ness and the Monster,* Newport (U.K.): J. Arthur Dixon, 1971 (superseded by Nicholas Witchell, *Loch Ness and the Monster,* Newport (U.K.): J. Arthur Dixon, 1975).

2. See David James, *Loch Ness Investigation: Annual Report,* London: Loch Ness Phenomena Investigation Bureau, 1969, p. 4.

3. J. A. Wheeler, "Drive the Pseudos Out of the Workshop of Science," *New York Review,* 17 May 1979, pp. 40–41.

4. Tim Dinsdale, *Loch Ness Monster,* London: Routledge and Kegan Paul, 1961, p. 234.

5. That case against the believers' evidence has been made most comprehensively by Binns *(The Loch Ness Mystery Solved),* Burton *(The Elusive Monster),* and Campbell *(The Evidence for the Loch Ness Monster).*

6. Gordon B. Corbet, "The Loch Ness Monster," *Nature,* 15 Jan. 1976, p. 75.

7. On the reported sensitivity of Nessies to sound, see Janet Bord and Colin Bord, *Alien Animals,* Harrisburg, Pa.: Stackpole Books, 1981,

p. 216; Daniel Cohen, *A Modern Look at Monsters,* New York: Dodd Mead, 1970, p. 109; Dinsdale, *Loch Ness Monster,* 1961, pp. 17, 37; Constance Whyte, *More than a Legend,* 3d rev. imp., London: Hamish Hamilton, 1961, p. 92; Rupert T. Gould, *The Loch Ness Monster and Others,* New York: University Books, 1969, pp. 82, 84.

8. For a recounting of St. Columba's story, see C. and O.G.S., *Antiquity* 8 (1934): 85–86; Dinsdale, *Loch Ness Monster,* 1961, pp. 33–37; Gould, *Loch Ness Monster,* 1969, p. 27; Whyte, *More than a Legend,* 1961, p. 135.

9. See F. W. Holiday, *The Great Orm of Loch Ness,* New York: Avon, 1970, pp. 174, 177; David James, *Loch Ness Investigation: Annual Report,* London: Loch Ness Phenomena Investigation Bureau, 1968; Robert H. Rines et al., "Search for the Loch Ness Monster," *Technology Review,* Mar./Apr. 1976, pp. 25–40; Charles W. Wyckoff, "Loch Ness and Underwater Photography," *Technology Review,* Dec. 1976, p. 50.

10. Gould, *Loch Ness Monster,* p. 114n4.

11. For example, see Henry Bauer, *Beyond Velikovsky: The History of a Public Controversy,* Urbana: University of Illinois Press, 1984.

12. On the inaccessibility of information, see Henry Bauer, "The Loch Ness Monster: Public Perception and the Evidence," *Cryptozoology,* no. 1, 1982, pp. 40–45; Henry Bauer, "The Loch Ness Monster: A Guide to the Literature," *Zetetic Scholar,* Dec. 1980, pp. 30–42; Henry Bauer, "Society and Scientific Anomalies: Common Knowledge about the Loch Ness Monster," *Journal of Scientific Exploration,* 1(1987): 51–74.

13. For example, see Elizabeth Montgomery Campbell and David Solomon, *The Search for Morag,* London: Tom Stacey, 1972, p. 74.

14. These are listed in Bauer in *Zetetic Scholar,* pp. 30–42. See also the "Books" section of the bibliography at the end of this book.

15. See Maurice Burton, *The Elusive Monster,* London: Rupert Hart-Davis, p. 72. I heard from Bernard Heuvelmans a few years ago that an Irvine film had been shown on television in Germany in 1981. Then, in the fall of 1984, about five seconds of such a film was included in Arthur C. Clarke's *World's Greatest Mysteries,* shown in America on ABC-TV. Subsequently I saw a longer version that had been shown on television in Britain in the early 1980s in a half-hour program on lake monsters, part of an Arthur C. Clarke series. It turned out to be the 1936 Irvine film, made by Scottish Film Productions Ltd. (1928), which is not at all convincing: nothing identifies the locale as Loch Ness, and throughout the disturbances caused in the water there are two dark objects, one fore and one aft, which do not change shape or elevation. Parts of the film were on view in 1985 at the Loch Ness Monster Exhibition, Drumnadrochit.

16. Ibid., pp. 48–49, 72.

17. See Whyte, *More than a Legend*, 1961, p. 4.

18. The incident is reported in Campbell and Solomon, *Search for Morag,* p. 107.

19. See John A. Keel, *Strange Creatures from Time and Space,* Greenwich, Conn.: Fawcett, 1970, p. 268; Peter Costello, *In Search of Lake Monsters,* London: Garnstone, 1974, p. 29; Nicholas Witchell, *The Loch Ness Story,* Lavenham (Suffolk): Terence Dalton, 1974, p. 28.

20. See Tim Dinsdale, *Monster Hunt,* Washington, D.C.: Acropolis, 1972, p. 57; Costello, *Lake Monsters,* pp. 26–27; Tim Dinsdale, *The Leviathans,* London: Routledge and Kegan Paul, 1966, pp. 57–87; Witchell, *Loch Ness Story,* 1974, pp. 27–28.

21. Costello, *Lake Monsters,* p. 28.

22. See Michael Enright, "Waiting for Nessie," *MacLean's,* 6 Sept. 1976, pp. 38–46; Nicholas Witchell, *Loch Ness Story,* rev. ed., London: Corgi, 1982, p. 65.

23. Whyte, *More than a Legend,* 1961, p. 7; Witchell, *Loch Ness Story,* 1982, p. 69. The *New York Times* reported on 21 Apr. 1934 (p. 9) that two of four photographs taken by the surgeon were to be reproduced in the *Daily Mail,* but only one was actually published. I have found no other reference to a second picture before Constance Whyte discovered it.

24. For reference to the computer enhancement, see Walter Sturdivant, "Loch Ness Monster," *European Community,* Apr./May 1976, pp. 34–40; Witchell, *Loch Ness Story,* 1974, p. 69; Steuart Campbell, "The Surgeon's Monster Hoax," *British Journal of Photography,* 20 Apr. 1984, pp. 402–10. In 1985 the computer-enhanced photo was on display at the Loch Ness Monster Exhibition, Drumnadrochit, and I could see nothing that would warrant a claim of whiskers on the jaw.

25. See Tim Dinsdale's works, *The Loch Ness Monster,* (2d ed., 1972; 3d ed., 1976), *The Leviathans, Monster Hunt,* and *The Story of the Loch Ness Monster*, London: Target, 1973; Roy P. Mackal, *The Monsters of Loch Ness,* Chicago: Swallow, 1976; and Witchell, *Loch Ness Story* (1974; 2d ed., 1976; rev. ed., Harmondsworth: Penguin, 1975).

26. Tim Dinsdale, "The Rines/Edgerton Picture," *Photographic Journal,* Apr. 1973, pp. 162–65; *Nessletter,* no. 12. I am not suggesting that the AAS was unduly secretive about the existence of a second "flipper" photograph but that the public perforce is misinformed about important developments. Mackal and the LNI were told of the second "flipper" in about 1973, and Mackal told me that a 1981 photo was mentioned at the first membership meeting of the International Society of Cryptozoology (New York, June 1983).

27. The photograph showing a tail-like structure is mentioned in Dinsdale, *Story of Loch Ness,* p. 112; Dinsdale, *Project Water Horse,* London: Routledge and Kegan Paul, 1975, p. 186; Witchell, *Loch Ness Story,* 1974, pp. 192–93. Roy Mackal explained to the author that "tail-like" was an early interpretation, later discarded, of the "two-body" photograph published in Rines et al., *Technology Review.*

28. Attempts at computer enhancement of the underwater shots are reported in Witchell, *Loch Ness Story,* 1976, p. 215; John Chesterman and

Michael Martin, "Return to Loch Ness," *Omni,* May 1979, pp. 92–95, 123–26.

29. See, for example, Dinsdale, *Loch Ness Monster,* 1972, pp. 120, 124, 127; Dinsdale in *Photographic Journal,* pp. 162–65; Dan Greenburg, "Japanese Come to Catch Loch Ness Monster," *Oui,* May 1974, pp. 50–52, 82, 112, 114–16, 118–22; Holiday, *Great Orm,* pp. 170, 211, 213; Witchell, *Loch Ness Story,* 1974, pp. 144, 165. In *The Monsters of Loch Ness,* Mackal published a comprehensive set of summaries of the LNI films, including stills from one of them—between nine and fourteen years after the films had been taken. The Loch Ness Monster Exhibition at Drumnadrochit gained access to the LNI files in 1985, and the exhibition's curator, A. G. Harmsworth, intends to make visuals available if the quality of the films permits.

30. See Mackal, *Monsters of Loch Ness,* Appendix D.

31. For mention of the hydrophonic data, see ibid., pp. 64, 76, 78. Until quite recently nothing had come of attempts to have the hydrophonic data interpreted; at last there are results, which Mackal told me in 1983 would be published in *Cryptozoology.*

32. The rejection by *Nature* is discussed in ibid., p. 32.

33. See Rines et al. in *Technology Review,* pp. 25–40.

34. For a discussion of the shortcomings of the Rines article, see Corbet, note 6.

35. The lack of published detail about the "flipper" photos and their enhancements left room for a charge that the published photos had been retouched. *Discover,* which had printed the charge, failed to publish the subsequent denial. The accusation was not repeated, however, in a later critique by the same people; other aspects of that critique have been answered. See Bernard Dixon, "Sea Serpent Survey," *Omni,* Sept. 1980, p. 18; *Zeit,* 31 Dec. 1976; Rikki Razdan and Alan Kielar, "Sonar and Photographic Searches for the Loch Ness Monster," *Skeptical Inquirer,* 9 (no. 2, Winter 1984–85): 147–56; Robert H. Rines et al., unpublished letter to the editor of *Skeptical Inquirer,* dated 15 Jan. 1985. See also subsequent published correspondence by Robert H. Rines et al., "Loch Ness Reanalysis: Rines Responds," *Skeptical Inquirer,* 9 (no. 4, Summer 1985): 382–87; Rikki Razdan and Alan Kielar, "Loch Ness Reanalysis: Authors Reply," *Skeptical Inquirer,* 9 (no. 4, Summer 1985): 387–89.

36. See Dinsdale, *Loch Ness Monster,* 1961, 111–13.

37. Dinsdale, *Monster Hunt,* p. 10.

38. Gould, *Loch Ness Monster,* 1969, pp. 12–13.

39. Whyte, *More than a Legend,* 1961, pp. xx, 37. For a discussion of how Nessie sighters are treated, see Burton, *Elusive Monster,* p. 11; Campbell and Solomon, *Search for Morag,* p. 68; Dinsdale, *Monster Hunt,* p. 58; Mackal, *Monsters of Loch Ness,* p. 17; Whyte, *More than a Legend,* 1961, pp. 3, 42, 68, 75, 119; Witchell, *Loch Ness Story,* 1974, pp. 128, 136; 1975, p. 97.

40. Strix, "Monster in Aspic," *Spectator,* 5 Apr. 1957, pp. 439–40.

See also Witchell, *Loch Ness Story,* 1974, p. 118; Skat Hoffmeyer, "Loch Ness—uhyret er mere en legende," *Aarhus Stiftstidende* (Denmark), 21 Apr. 1957.

41. For example, see Dinsdale, *Loch Ness Monster,* 1961, p. 13; Campbell and Solomon, *Search for Morag,* pp. 32–33; Gould, *Loch Ness Monster,* 1969, pp. 14–15; Whyte, *More than a Legend,* 1961, p. 79.

42. The desire to have science accept lay observations is expressed in William B. Corliss, *Mysteries Beneath the Sea,* New York: Thomas Y. Crowell, 1970, p. 139; Dinsdale, *Project Water Horse,* pp. 183–84; Witchell, *Loch Ness Story,* 1974, p. 79.

43. Although the monster hunters may share the belief that Nessies are real (and even that belief is held to varying degrees), each monster hunter has an idiosyncratic notion about what they are and assesses the evidence accordingly. Dinsdale, for example, thinks that the animals are probably related to plesiosaurs, and so he puts emphasis on the long and muscular necks that have been reported. Mackal, by contrast, regards the animals as more likely to be mammals and so is skeptical of reports of really long necks. Those differing assessments illustrate Kuhn's much-discussed account of the progress of scientific knowledge: each person's prior paradigmatic view determines the weight he is willing to accord to any given fact; different people ascribe different weights therefore, and disagreements about "the facts" are inevitable. See Thomas S. Kuhn, *The Structure of Scientific Revolutions,* Chicago: University of Chicago Press, 1970.

44. For the Gray photo, see Gould, *Loch Ness Monster,* 1969, p. 23; E. D. Baumann, *The Loch Ness Monster,* London: Franklin Watts, 1972; Ellen Rabinovich, *The Loch Ness Monster,* New York: Franklin Watts, 1979; George R. Zug, "Once More into the Loch," *Encyclopedia Britannica Yearbook of Science and the Future,* 1978, pp. 154–69; Burton, *Elusive Monster,* p. 79. For the Irvine films, see Costello, *Lake Monsters,* pp. 71–72; Mackal, *Monsters of Loch Ness,* p. 290. For the surgeon's photo, see ibid., pp. 96–98; Burton, "The Loch Ness Monster," *Illustrated London News,* 20 Feb. 1960, p. 316; Burton, *Elusive Monster,* pp. 159–65; Burton, "Verdict on Nessie," *New Scientist,* 23 Jan. 1969; Campbell in *British Journal of Photography,* pp. 402–10; Dinsdale, *Loch Ness Monster,* 1982, pp. 200–201; Richard Whittington-Egan, "Loch Ness: The Monstrous Zoological Problem," *Contemporary Review,* 225 (Sept. 1974): 138–44. For the Macnab photo, see Mackal, *Monsters of Loch Ness,* pp. 273–76; Campbell and Solomon, *Search for Morag;* Costello, *Lake Monsters;* Dinsdale, *Loch Ness Monster,* 1982, pp. 204–5; Whyte, *More than a Legend,* 1957; Witchell, *Loch Ness Story,* 1974.

45. Dinsdale has told me that his subsequent omission of the photo resulted from O'Connor's attitude: upset by public criticism and ridicule and charges of fakery, O'Connor wanted to be left in peace. Dinsdale re-

mains inclined to believe that the photo could really be of a Nessie.

46. For the O'Connor photo, see Costello, *Lake Monsters;* Rabinovich, *Loch Ness Monster;* Mackal, *Monsters of Loch Ness;* Burton, "Verdict on Nessie." For the Cockrell photo, see Burton, *Elusive Monster;* Dinsdale, *Loch Ness Monster,* 1961; Dinsdale, *Story of Loch Ness;* Witchell, *Loch Ness Story,* 1974; Mackal, *Monsters of Loch Ness.* For the unattributed photo, see Burton, *Elusive Monster,* p. 80; Costello, *Lake Monsters,* p. 64; Mackal, *Monsters of Loch Ness,* p. 99; Witchell, *Loch Ness Story,* 1974, p. 51. For the Fraser still, see Burton, *Elusive Monster,* pp. 48, 72; Palle Vibe, *Gaden I Loch Ness,* Denmark: Rhodos, 1970; Costello, *Lake Monsters.*

In 1981 I inquired into the provenance of the latter photograph. The author of the book in which it had appeared could no longer remember it and referred me to the copyright holders, Nordisk Pressefoto. The latter ignored my request for details about the picture and responded that the foreign copyright was held by Keystone Press, which wrote that "the photograph in question was sent to us by Nordisk Pressefoto. . . . It was taken several years ago by one of their men while he was in Scotland. I do suggest you write direct to them. . . ." Which I again did; reply: "We regret not being able to give you further information of the photographer James Fraser as we don't have the original photo in our files any more."

47. See Bauer in *Zetetic Scholar,* note 12.

48. Burton's reasoning is discussed in chap. 7. My analysis of Binns's book is in *Nessletter,* no. 70, June 1985.

49. See Whyte, *More than a Legend,* 1961, pp. 7, 10; Witchell, *Loch Ness Story,* 1974, pp. 66–69.

50. See Campbell and Solomon, *Search for Morag,* pp. 130–34.

51. The postcards are discussed in Gould, *Loch Ness Monster,* 1969, p. 100. Larry Laudan reminded me that it is not uncommon that literature intended for tourists is unreliable!

52. "Inverness Two in Top 20," *Highland News,* 26 Apr. 1985. Glasgow's Burrell Collection Gallery led with 1.1 million visitors, Edinburgh Castle was second with 847,000; the monster exhibition was nineteenth, below the Inverness Museum and Art Gallery, which was seventeenth with 173,000 visitors.

53. For a discussion of land sightings, see Costello, *Lake Monsters,* p. 333; Dinsdale, *Loch Ness Monster,* 1961, p. 30; Gould, *Loch Ness Monster,* 1969, pp. 43, 50; Mackal, *Monsters of Loch Ness,* appendix A, table 2; Burton, *Elusive Monster.*

54. For the July 1933 sighting, see Dinsdale, *Leviathans,* pp. 55–57; Gould, *Loch Ness Monster,* 1969, pp. 43, 50, 86, 87, 157. The January 1934 sighting is discussed in Dinsdale, *Loch Ness Monster,* 1961, p. 45; Whyte, *More than a Legend,* 1961, pp. 73–76.

55. See Dinsdale, *Loch Ness Monster,* 1961, p. 39; Dinsdale, *Story of Loch Ness,* pp. 19–24; Gould, *Loch Ness Monster,* 1969, p. 27; Holiday,

Great Orm, pp. 100–109, 131–48, 181–96; Whyte, *More than a Legend,* 1961, pp. 125–46, 158–60.

56. See Dinsdale, *Loch Ness Monster,* 1961, pp. 169–82; Whyte, *More than a Legend,* 1961, pp. 137–38, 146–51. This seems an appropriate place to reemphasize my disclaimer: As I describe dilemmas inherent in Nessiedom, it may seem that I am criticizing. Not so. Constance Whyte was right to give this logically necessary portion of the story; nevertheless it fuels the skepticism of those who are not inclined to believe.

57. For stories of other such animals in Scotland, see Dinsdale, *Loch Ness Monster,* 1961, pp. 38–40, 169–82; Whyte, *More than a Legend,* 1961, pp. 123–33, 136.

58. Dinsdale, *Monster Hunt,* pp. 36–37; Holiday, *Great Orm,* p. 179; F. W. Holiday, *The Dragon and the Disc,* New York: W. W. Norton, 1973, pp. 48–79.

59. Holiday, *Great Orm,* p. 125; Mackal, *Monsters of Loch Ness,* p. 117.

60. See Peter Lancaster Brown, *Megaliths and Masterminds,* New York: Charles Scribner's Sons, 1979.

61. Holiday is not the only one to suggest that monsters might be aware of cameras and be able to avoid them. See Bord and Bord, *Alien Animals,* p. 36.

6

The Quest

By 1972 or 1973 I was more inclined to believe than to disbelieve that Nessies exist. I had read all the available books on the subject and much of the more scattered material; and I had been for a number of years an overseas member of the Loch Ness Investigation Bureau (LNI), regularly receiving the bulletins and newsletters. Joking only a little, I would tell my friends that my knowledge of Nessie had passed the standard scientific test of permitting accurate predictions to be made: when in April 1972 the newspapers announced that a dead Nessie had been found floating in the loch, I had been able to diagnose this immediately as a hoax because I knew roughly what a Nessie looked like, and the published description did not fit! Sure enough, the next day my diagnosis was confirmed.[1]

I paid a visit to Loch Ness in the spring of 1973. At the tourist bureau in Inverness I was startled to see several photographs of Nessie that were entirely new to me; moreover, at least one of them looked wrong—the head, in profile, was too big and not quite the right shape (see fig. 17). A man behind the counter assured me that the photographer, Frank Searle, was well known locally and accepted as a genuine and dedicated monster hunter. In some haste I drove to Searle's camp and was stunned by his display of photographs: in the space of a year or so he had obtained more and better photographs than had the LNI in ten years of teamwork, more than had accumulated from all other sources in the preceding forty years. Searle himself made me a little uneasy, and several of the pictured shapes looked subtly, indefinably, but definitely wrong—but who was I to put my judgment above that of locals who had been in contact with the man for nearly four years? Besides, I was not really so sure that I knew exactly how Nessies looked.

By the greatest good fortune a way to resolve my dilemma came almost immediately to hand. The LNI was no longer active at the loch, but I chanced on one of its more prominent members, introduced myself (university professor, fellow member of LNI), and expressed my doubts about that most peculiar of Searle's pictures. The response was evasive: "a very *important* picture," I was told, in a manner that refused to give it an actual imprimatur of validity yet equally definitely refused to cast overt doubt on it. I was troubled and confused and taken aback, and certainly no wiser. But I have come to understand much since that time and long ago forgave my now dear friend his reticence on that occasion. I recount this personal experience now because it touches on dilemmas in which the monster hunters find themselves: the credibility of their enterprise is jeopardized by the very actions that they take to safeguard it, and that again is inevitable given the nature of the quest.

SCIENCE AND NESSIEDOM

Such dilemmas can best be described, I believe, by contemplating their absence in research that is within the accepted bounds of science. One becomes potentially a scientist after lengthy training and upon the display of a modicum of relevant competence and of such other desirable qualities as honesty and reliability. At first, one's place in the profession is probationary; one must demonstrate the ability to put into effective practice what one has learned: the knack of finding problems to solve that are neither too trivial nor too intractable; the discipline to prove one's point beyond reasonable doubt through weary repetitions and self-criticism; a capability to communicate one's work orally and in writing; and other things besides. If one can do all that, and has good luck as well, one becomes settled in the profession, with tenure in a university or its equivalent in other organizations. But freedom in one's work is by no means absolute even then. A successful career is made through respecting an untold number of largely implicit and tacitly accepted rules; and transgressing them not only blocks advancement but can even bring dismissal from the most securely tenured post—dishonesty in one's work, for example.

The community of science disciplines itself quite rigorously, and the attaining of high status within that community bespeaks real accomplishments. The wider society knows this and consequently accepts as reliable the pronouncements made by prominent members of the scientific community and by those who speak for recognized associations of scientists. The wider society also has its ways of discouraging those who would attack or compete with the scientific establishment: on the whole, for example, the orthodox practitioners of medicine have had the support of government in keeping homeopaths at bay; and the American Chemical Society need not fear that an alchemical association might gain the authority to approve curricula; and the biologists have been largely successful in resisting the political maneuvers of the creationists. The confidence that society has in the profession of science permits the latter to go about its internal affairs without fuss. Individual transgressions that could become scandals—fraud, for example—can usually be settled without the dirty linen being exposed to public scrutiny and derision.

Thus scientific activity is carried on by people who are well schooled and who respect a host of tacit guidelines. One uses methods and apparatus that have been fully tried and tested and described in literature available to everyone. If one needs to use a new approach, one develops and tests and publishes it. One acquires the ability to judge what ought to be published, when it is ready for publication, and where to publish it. One knows the subtleties defining what may be kept secretly to oneself for a while, what can be discussed at a meeting, and what must be committed to writing. One knows that publication in a professional journal must precede announcements in the press. There is a continual exchange of information with others working in roughly the same field, and there are tacit rules to limit the extent of the unavoidable competition: scientists do not boldly enter a field tended by others and preempt their results. Manuscripts for publication are reviewed most stringently, so that what finally appears meets the standards of the profession in the reliability of the data, the cogency of the reasoning, the inherent value of the report, the giving of appropriate credit to others through citation of their work.

In consequence, the scientific literature is progressive and cumulative; the store of certified knowledge expands, the later writings on a particular subject are more informative and definitive than the earlier ones. The wider society is impressed by this and supports much basic scientific work largely because of a perceived connection between scientific research and higher education and technology that benefits society. And because of the demonstrated trustworthiness of the profession, the assigning of support is largely left to the profession itself. The training of scientists and the tacit rules under which they work ensure that what finally becomes science is more reliable, coherent, valuable than could be achieved by people, no matter how brilliant individually, working in isolation and without the generally agreed rules. Through discipline, science transcends the failings and weaknesses of individual scientists.

The search for Nessie, carried on outside the establishment of science, offers none of these manifold advantages and conveniences and perquisites. Anyone can set up as a monster hunter, without any training and irrespective of appropriate competence or ability (in point of fact, we do not even know what the appropriate abilities and competences are). One has "tenure" at will; there is no probation, no test to pass, no standard to be continually met, no mechanism for weeding out the incompetent or fraudulent. There is no structured hierarchy through which the best become generally respected and acknowledged by the wider society, which then looks to those people as the authoritative spokesmen for the field. When there are disagreements within Nessiedom on matters of substance or ethics, there is no way of resolving those internally, and the unwashed linens are publicly displayed—not always as dramatically as in August 1983, however, when a petrol bomb was thrown at the boat of the Loch Ness and Morar Project.[2] Whereas science transcends the imperfections of its practitioners, Nessiedom displays the shortcomings of the individual hunters; and that serves to discredit the quest in the view of the conventional wisdom and to make belief in Nessies that much harder to achieve.

THE MONSTER HUNTERS

That Nessiedom is not science has implications for who the monster hunters are, how they relate to one another and to outsiders, and how they are viewed by the public media. Since Nessiedom lacks the organization of science, inevitably it appears unsatisfactory from the viewpoint of the scientists, whom the hunters are seeking to convince and enlist. The hunters work under a number of pressures of which scientists are not aware, and they are always at the mercy of a possible and ironic unfairness: the "best" hunters may not get the best results. In science there are high correlations among ability, training, achievements, and reputation; those who rank high in such characteristics reasonably expect to continue to make significant achievements. It is very unlikely that a graduate student will snatch a Nobel prize from a senior scientist, still more unlikely that some nonscientist will do so.

In Nessiedom, in part because it is exploration as much as experimentation, luck and chance play a huge and possibly even decisive role. Indeed, it is entirely possible that the definitive proof of Nessie's existence will come from some naive tourist, entirely ignorant of the history of the quest, perhaps even equipped with ridiculously expensive gadgetry which he doesn't know how to use properly but which nevertheless does the job. Were I a dedicated hunter over these many years, scrambling for funds to implement excellent ideas, always frustrated by slow progress, I would occasionally have nightmares about that possible irony and about the even worse possibility that one of the charlatans might have the luck—one who has previously faked photos might nevertheless be in the right place at the right time with the right equipment.

Among the monster hunters one finds individuals of the highest ability and character. Gould and Whyte epitomize the dedicated amateur, with energy and common sense, wide culture and high education. Dinsdale's competence in engineering and in photography are linked with intense intellectual curiosity and a deep sense of ethics, the ethics appropriate to truth seeking as well as the much more difficult practical ethics of dealing with other people. The teams organized by Robert Rines offer a dazzling combination of the highest expertise, in photography generally and underwater photography in particular,[3] in sonar, and in electronic gadgetry in general. Their work should, in time, come to be seen as a model of how one attacks a novel problem—a careful sifting and accumulation of data (looking for correlations among sightings, sonar mapping of the bottom of the loch), inspired generation of ideas and hypotheses, determined testing and perseverance, much error but always new trials.

Many other people with similarly admirable attributes have been part of the quest at various times. Yet even people of such quality have been denigrated and ridiculed,[4] in part because their credentials are not in biology; in part because the conventional wisdom ridicules the quest itself; and in part because the Dinsdales and Rineses are seen by the public as belonging to the general class of monster hunters, which does include also some charlatans and some who dabble in pseudoscience and the occult (see chap. 6). Some maverick biologists (see chap. 7) suffer their own peculiar dilemmas when they associate themselves with the quest.

Inevitably an outside observer finds it easy to ascribe to the whole heterogeneous group the doubtful or crankish characteristics of some of the individuals. The absence of a guild, an established professional association, makes it impossible for the genuine monster hunters, the dedicated truth seekers and people of integrity, to disassociate themselves from the others. Within science there is a tacit discipline that sets limits beyond which it becomes self-destructive for a scientist to act on the less admirable motives; and the structure of science offers tangible rewards to those who sublimate direct and inordinate self-interest into relatively disinterested and lawful inquiry. In Nessiedom, neither this stick nor the carrot apply. One who seeks personal advantage from the quest can and must do so in the wider society directly, competing with fellow monster hunters to be regarded by the press and the public as the best, the first, the most admirable.

To the informed student it is clear, for example, that Tim Dinsdale's knowledge of the whole business is second to none, his books and public statements reliable and principled; and that the Academy of Applied Science has brought the ingenuity and technical sophistication of the search to the frontiers of available technology. In science that would ensure that others interested in the quest would take pains to seek the advice or approval of these established workers. Not so in Nessiedom. The Loch Morar Project, for reasons not clearly stated, became the Loch Ness and Morar Project (LN&MP), canvassing widely for members and funds, stating (rather wishfully) that "[we are] extending our influence and have become, in the eyes of more and more people, the authority on the controversial issue of the Loch Ness Monster."[5] The project[6] was planning a public relations center at Loch Ness, which would then have given the visitor four such centers to choose from: one at the Drumnadrochit Hotel, the most reliable and with full information about the most recent results, supported by Dinsdale and Rines; one at Fort Augustus, which mentions some of that work but devotes about one-quarter of its space to the LN&MP, which as of 1980 had done little more than repeat and extend the AAS sonar studies at Loch Ness; Searle at Foyers; and the proposed LN&MP center. Public confusion, bad enough already, could thus only increase and thereby hinder rather than help the quest itself.[7] Already in 1976 the *Inverness Courier*, in an editorial on 14 June, had drawn attention to the unfortunate council actions denying a site to the LNI but granting one to Searle.

The public simply does not know who the real authorities on Nessie are, in part because a portion of the monster-hunting community is engaged in public, political jockeying for these positions instead of allowing demonstrated achievements to determine them. The genuine hunters cannot disallow, nor effectively in the public view disassociate themselves from, such stunts as the Japanese expedition of 1973, or the exorcism of Nessies, or the use of psychic powers to bring Nessies to the surface for photographing, or the activities of Searle.[8]

Some of the most dedicated searchers refrain from public criticism of their unwanted fellow hunters. In her description of self-interested motives (see below), Campbell gives no names, though her words are surely based on real and not fictional characters; so too my friend's reticence about Searle's photograph before he came to know me. In part the eschewing of public criticism results from sheer personal distaste for it, in part from the fact that there are better things to do with one's time in furthering the quest. But, I believe, it is also partly because the wider society has such a lack of respect for the whole venture of Nessie seeking: there is the fear that any acknowledgment of dishonesty or unethical behavior on the part of anyone who claims to be a hunter might further damage the credibility of the central matter. (This is a problem in anomalous fields in general. Certainly the parapsychologists feel that they are not given credit for themselves uncovering fraud—Walter Levy in Rhine's institute, for example—but rather find their whole field denigrated because of an alleged prevalence of fraud.)

Inevitably, then, there is a tendency to close ranks against the outsiders and for very strange bedfellows, from competing groups and antagonisms, to coexist in Nessiedom without public disagreement. The insiders have their own knowledge of who is a fraud and who is self-serving, of the strengths and weaknesses of their bedfellows, but they are reluctant to share that knowledge with the wider society. That laudable and understandable reluctance, however, also helps to bring about what it is designed to avoid; for example, it allows the local council to permit Searle a site for his information center, the newspapers to publish his pictures, and Searle to be for some people the only known monster hunter,[9] all of which can enormously damage the public's view of the hunt once it becomes widely enough recognized that the man is a fraud.

As a student of chemistry I learned not only about chemistry but also—and largely implicitly—certain professional ideals: primarily that one should be moved only by the passion for new knowledge objectively established and not by personal self-interest in the practice of one's profession. Through subsequent experience I learned that practicing scientists are humanly prone to self-interestedness and other human frailties. Perhaps, as with priests, their high ideals make their practice more ethical than it would be in the absence of those ideals, but the practice is certainly not always the ideal. When I became interested in the quest for Nessie I began with the naive notion that the Nessie hunters were moved only by the desire to uncover the truth: since Nessie hunting was perforce an avocation or hobby and not a way to earn a living, the motive of curiosity would not be tainted by careerist motives or the like. That was indeed naive, for the story of the quest contains many individual histories of mixed motives.

Elizabeth Montgomery Campbell has pointed out that, coexisting with the sheer curiosity about these remarkable animals, "There are . . . other less disinterested motives. . . . There is the very natural desire to be proved right, or at least to bring about a triumph of open-mindedness over prejudice. . . . Perhaps there is the desire to make one's mark and to go down in history as one of the great discoverers. For some people the life style associated with investigations of this type becomes almost an end in itself, tending to obscure the original objective."[10] There are doubtless some hunters for whom it is essential not that Nessies be properly discovered but that they be the ones who do the discovering; some who seek personal vindication through success, perhaps vindication for the absence of a conventional career or against the Establishment that rejected or mocked them. All human qualities are on display in Nessie hunting, as in other human pursuits.

So we should not expect the hunters to be purer than others, nor should the quest itself be criticized because it is carried on by human beings. I suspect, however, that the quest and the hunters often are judged by comparisons with the ideal, something we tend to do frequently. We talk as though we expect all doctors to be exemplars of the Hippocratic Oath, as though we expect the law to infallibly deliver justice, as though we expect politicians to be statesmen. In conventional pursuits we are often willing to forgive certain lapses if the goal is sufficiently important: lawyers' tactics are not criticized if the right verdict is reached, stealing of information by reporters is applauded if the public good seems to be served thereby, and so on. But the quest for Nessie is not universally or even widely seen as an important endeavor, and so perhaps we expect the monster hunters to behave more impeccably than do most other groups of people. Moreover, some of the Nessie hunters encourage such expectations by describing the quest as a search for truth in the face of an indifferent or antagonistic science that ought to be searching for truth: they virtually invite comparisons of their quest with the ideal.

Individual monster hunters display mixtures of characteristics that one finds in all humans, but the public image of individual monster hunters is tainted by the aura of amateurism and the suspicion of fraud that permeates the quest. The idiosyncracies of scientists of great repute are usually played down by the media because of the awe in which science is held; by contrast, much is made of the human failings of individual monster hunters because the quest itself is held in scant respect.

ORGANIZING THE HUNT

From the first wave of publicity in 1933 until 1960, most of the evidence for the existence of Nessies had come by chance or by luck. Gould in 1934 and Whyte in 1957 had brought together eyewitness reports and the few photographs obtained serendipitously and had made excellent cases through logical and incisive analysis; Dinsdale in 1960 had the good fortune to film a Nessie during the single week that he was able to spend at the loch. But it was rather clear from the beginning that such infrequently surfacing creatures required that the loch be under observation by well-organized groups if there was to be a decent probability of confirming Nessie's existence. Sir Edward Mountain in 1934 hired twenty men who stood watch ten hours a day for five weeks; the yield was about twenty reported sightings and five printable photographs, only one of the latter not being explicable as the wake of a boat.[11] Captain James Fraser continued the watch for several weeks more and shot a 16mm film of a large object disturbing the water; but there was not enough detail in the film to persuade the scientific community that this was no seal or otter.

To bring Nessie into science calls for incontrovertibly clear, detailed photography or for the acquisition of a specimen. The habitat is more than twenty miles long, a mile wide, and hundreds of feet deep, with water that absorbs light rather as does weak tea and thereby limits visibility to twenty or thirty feet even under powerful illumination. All the clever ideas that continue to be suggested require large sums of money or large numbers of people or both if the chances for success are to be reasonably good. The last two decades of the quest provide examples of heroic organizational efforts to meet these requirements; and the continuing lack of substantive success illustrates the difficulties such efforts encounter, in part because the imprimatur of science has not been granted.

In the early 1960s the Loch Ness Phenomena Investigation Bureau Limited (LNPIB, later shortened to LNI) was founded, initially to continue Constance Whyte's endeavor of recording and analyzing sightings. Soon the group began to organize expeditions with David James, a member of Parliament with a distinguished record of military service in World War II, as the driving force. Such an organization required funds and manpower, so the organizers canvassed for members. The LNI, over its decade of activity, grew to a membership of more than 1,000. Its history shows, however, that the manpower and funds thus generated could not sustain such expeditions. Most of the members were not in a position to travel to Loch Ness and participate directly. Instead they had to be kept interested through newsletters, annual reports, and responses to individual letters, which involved much expenditure of time and used amounts of money comparable to that generated by the membership, leaving totally inadequate funds to support the hunt itself.

Members did provide volunteered time to staff the camera sites LNI set up for varying periods between 1963 and 1973, but to a certain extent the dictum applied, "You get what you pay for": the volunteers were enthusiastic, but training and self-discipline were uneven. For example, a presumed sighting that persisted for about twenty-five minutes at very long range produced only fifteen seconds of film through the long telephoto lenses, presumably because the viewfinding arrangement—very small field of view—had not been mastered by the operators. The long hours of watching sometimes sapped concentration and devotion to duty; for instance, one sighting failed to produce film because the man and woman supposedly at the camera were otherwise occupied. Over the years the LNI extended its efforts until more than 70 percent of the loch's surface was under its surveillance. Yet such is the role of luck that most of the accredited sightings continued to be made by chance by tourists or locals, not by members of LNI teams—in 1968 only two of fourteen accepted sightings came from LNI teams; in 1969 only three of fourteen.

Naturally David James looked wherever possible for additional support: to private patrons and newspapers and television companies for funds, to all sorts of people for expertise and equipment. But just as with the LNI membership, every potential or actual source of support brought with it some liabilities. Deluged with correspondence, James nevertheless responded to the most unlikely sources on the off chance that they could lead to useful help: youngsters wanting material for a school project received full replies in case they had wealthy relatives who just might become interested; self-interested inventors were treated courteously just in case—one saw Loch Ness as a way to advertise his expertise with submarines, another with lighter-than-air vehicles.

The early effort was supported by Associated Television, and it seemed reasonable in exchange to grant that company the rights of distribution of television coverage. But later, when much more substantial funds were provided by Field Enterprises of Chicago, that earlier guarantee made the negotiations more difficult; the issue of rights became quite complicated and wearying to the Nessie hunters. When Jacques Cousteau's Institut Océanographique made overtures to LNI, the latter had to bring Field Enterprises into the discussions. When electronics experts from the University of Birmingham collaborated with LNI in pioneering work with sonar, they could not just get together and tangle with the technical challenge, they had to evolve documents safeguarding the investigators' rights to publish their results first in a technical journal and Field's rights to subsequent distribution in the popular media.

The task of finding out what Nessies are is typical of investigations in what is called "pure" science. There is a reasonably well formulated goal, but one does not know beforehand what method will succeed, how many barriers and blind alleys remain to be negotiated, how much time and how many resources will be needed; in fact, whether the goal is attainable. Those who support pure science learn not to ask for definite results in a given time, to give the investigators the maximum autonomy, and to demand only that the work proceed competently and diligently. But the quest for Nessie does not have the acknowledged status of science, and such groups as LNI must accept support from sources that have their own vested interests.

Field Enterprises made it possible for LNI to work on a scale and with equipment that would have been impossible without their help; but Field had the self-interest of using the quest to publicize its *World Book Encyclopedia* and wanted definitive results fairly quickly. That led inevitably to friction and to dilemmas and distractions for LNI: for example, Field imposed on David James and his colleagues to entertain some individuals who had won a trip to Loch Ness in competitions sponsored by Field Enterprises. And at times Field gave the impression of believing that it "owned" Nessie. The National Geographic Society had devoted one of its television specials to Loch Ness and included an interview with one of the LNI workers. When that show was broadcast in the United States, Field was annoyed because the sponsor of the program happened to be the publishers of *Encyclopaedia Britannica*, their competitor. Even beyond those dilemmas was the fact that some British pundits criticized LNI for allowing control of the quest to pass, allegedly, into the hands of commercial American interests.[12]

These strings attached to sources of funds for LNI[13] have seemed even more irksome because to the Nessie hunters the level of support corresponded neither to the required effort nor to the potential profits to the sponsors if the quest succeeded. Field Enterprises was the largest single contributor to LNI, providing more than $100,000 over four years. Yet that amount represented probably no more than half of LNI's expenditures, certainly much less than half if the monetary value of the volunteers' time were considered. In exchange for its support, however, Field had essentially sole rights to commercial exploitation of results (though LNI was guaranteed a share of profits from the exercising of those rights).

The failure of sponsors to recognize value received from LNI in exchange for their contributions must have appeared particularly perverse in the case of local and regional Scottish authorities. Only once in the decade of LNI's work did the Highlands Development Board make a donation, 1,000 pounds. The LNI was permitted to use a site—privately owned—overlooking the loch; but there were continual demands that the site be improved, and ultimately LNI lost permission to continue its search from that base. Yet there was clear evidence that LNI had served the interests of large numbers of tourists. The tally of visitors to the LNI information center grew from about 1,000 in 1965 to 25,000 in 1966, then more slowly to 54,000 annually by 1971. The failure of local authorities and businesses to make possible a continued existence for LNI might indicate, by the way, that the Loch Ness monster is hardly a creation of the tourist industry.

Just as the LNI might have resented the quid pro quo exacted by sponsors, so individual members of LNI might have chafed under the rudimentary discipline necessarily imposed by that organization. The watchers at Loch Ness received no salary, of course, and they paid at least some of the cost of their food, though accommodations were furnished without charge; yet they had to relinquish any rights to results they might obtain: "All participants must sign a copyright form so that the rights in any material obtained are vested in the Bureau."[14] Again, in the LN&MP, ". . . a field agreement . . . must be signed which makes all photographs taken in the field copyright of the Project and prohibits the writing of articles or interviews about the work of the Project."[15]

Strains of these sorts make it unlikely that such an organization as LNI could have a long life. Members are loyal to it for only one reason: they want it proved that Nessies exist. By contrast, a scientific association or society serves its members in several ways: by affirming their professional standing and credentials; by providing avenues for publication, for presenting papers, for meeting people who can be useful to one's career; by acting as a trade union and using political means to bring benefits to members. So chemists, for example, who see as misguided some action or pronouncement by the directors of the American Chemical Society, usually remain paying members nevertheless because of the many compensating advantages. But with LNI and its ilk the pursuit of the quest is the only cementing bond. If newsletters are not prompt and informative, or if some members believe that other tactics should be used, or for any number of other petty reasons, the organization is vulnerable to a loss of members, or to schisms in which other groups form and compete for support from prospective patrons.

In science, neophytes learn their business under the tutelage of experienced practitioners: the institutions of science ensure that both those who learn and those who teach benefit from the interaction, in the short term and over the long run. In Nessiedom, by contrast, there is no incentive to become an apprentice or to take apprentices, so each new hunter or group of hunters makes something of a fresh start, not necessarily benefitting from what earlier workers have done. Dinsdale's writings over the years make plain how much he has learned since 1960. The Academy of Applied Science has also learned, beginning in the 1970s, especially about the tribulations of interacting with sponsors and the public media. The Loch Ness and Morar Project is now beginning to lose some of the impertinently aggressive stance with which it began around 1980; and the exhibition at Drumnadrochit has moved from the naive advocacy of 1980 to a more judicious interpretive stance. In theory one might suggest that those who joined the quest later could have learned more from those who had carried it on earlier; in practice, however, that does not happen because the apprenticeships common in professional science cannot be established in ventures that are carried on by people using their vacation time and personal funds.

Within science, competition among research groups takes place in an orderly manner prescribed by long-standing traditions; and the progress of science results as much from those traditions as from the individual brilliance of scientists. In Nessiedom, however, there is no such governing tradition of how research ought to be done or of the permissible limits of competition. So there is jockeying for the public limelight, and research is often repetitive rather than mutually reinforcing and cumulative. The Plessey Company did sonar work at the same time as the group from the University of Birmingham and may have repelled Nessies because it used such a low frequency of sound, a clearly audible ten kilohertz.[16] The National Geographic Society spurned Rines's photos, thinking their own crew could do better, and set up their apparatus right next to Rines—with total lack of success and perhaps also discouraging the creatures from venturing, as they had the previous year (see figs. 8 and 9 in chap. 2), near Rines's cameras.[17] Eyewitnesses are not only subject to ridicule by the media but are plagued also by an endless succession of independent private investigators asking them to repeat their stories.

DEFICIENCIES AND DILEMMAS

Given the lack of organizational structure and the absence of a tradition of disciplined inquiry, it follows that the monster hunters' activities are open to easy denigration when viewed by the standards of science. Discoveries are announced before they are assured, for example. Thus stonework found underwater in the loch was immediately interpreted as megalithic, related to Stonehenge and so forth, but shortly thereafter it was realized that the stones were merely dumps from work on the Caledonian Canal.[18] Cancellation of the Edinburgh symposium because the new pictures of 1975 had already been mentioned in the press may have been an excessively conservative act by the Establishment, but it was hardly in the scientific tradition that, after the photographs had been made available to the British Museum by the AAS, the latter realized that further calibration tests at Loch Ness were needed. The museum ought never to have been in the position of being able to make a statement—that the results were inconclusive—before it had all the technical details and calibrations, which should have accompanied or preceded the photos.[19]

The difficulty of obtaining photographic results may also tempt enthusiasts to push interpretation of the data farther than caution would suggest. Thus the quite detailed report on the underwater photographs of the 1970s provides grist for the skeptical mills at several points:[20] in estimating distance by characterizing as "well in focus" a portion of a very indistinct shot; in offering dimensions for an indistinctly illuminated object in another shot; in claiming a resemblance between one totally indistinct shape and another slightly more definite one; in comparing the separation of the "horns" on the "head" (see fig. 9 in chap. 2) with that of wakes filmed at long range. On the one hand it is natural for the hunters to venture such speculative conclusions and may be heuristic for their own purposes; on the other hand making such speculation public gives the debunkers easy ammunition when they criticize an apparent lack of scientific rigor.

In somewhat that same vein we note instances where understandable naivety on the part of the monster hunters damages their credibility. For example, the exhibition at the Drumnadrochit Hotel (in 1983 and since the summer of 1980) claimed proof of a certain point when that proof depended on photographs yet to be received. Hardly more convincing than the statement that the Loch Ness Investigation Bureau ". . . gathered so much data and so many accounts, that to show it all would take up more space than we have for the whole exhibition"—a claim unlikely to disarm the skeptic, particularly the knowledgeable skeptic who was aware that ten years of effort by the LNI had failed to produce scientifically compelling evidence. Inevitably unconvincing are the attempts to convince others only on the basis of the strength of one's own personal conviction—Gould on the reliability of the witnesses he interviewed, for example; or Dinsdale's assurance that we can trust a witness he had good but undisclosed reason not to name, or his asking us to believe that he was advised to keep his film secret, for no stated reason, by someone he does not name.[21]

The cumulative effect of trivia of this sort further hinders the acceptance of the quest as respectable by the scientific establishment and the wider conventional wisdom. Yet all this is inherent in the nature of the enterprise. For instance, when the hunters seek to publish in mainstream journals their contributions are rejected; when biologists publicly join the hunt they do so at their peril; and the media feel free to smear the most genuine of the hunters unconscionably, as they would not feel free to do with scientists who are authorities in an established field and backed by professional organizations of long standing. Even when prominent people set up an organization to lend some respectability to the venture the effect is not great: the patrons of the LNI included well-known naturalists and the organizer was a member of Parliament, yet five years after the group was formed its existence was not well known to the media.[22]

Thus hampered in attempts to establish connections with science, inevitably the venture finds itself connected instead with other fringe groups and subjects. Over the years a number of attempts have been made to interest the community of biologists: Mountain's film was shown to the Linnean Society of London, and evidence was submitted without publicity on several occasions to panels of experts.[23] But finding no interest aroused, inevitably the hunters have displayed their data in the public media and in publications that bear the taint of crankishness.

One further source of difficult relations between science and Nessiedom has to do with control of the data. In science it is expected that one will share all of one's data with others through open publication; the discoverer, however, is guaranteed permanent credit because of prior authorship. The monster hunters do not have available to them this means of both communicating openly and yet also preserving ownership rights. Consequently, they are reluctant to put all their hard-won results into the public domain, to be used by others ad lib, even by popular magazines in their never-ending campaigns for larger circulations. That reluctance in itself exacerbates the tension between the hunters, the Establishment, and the media: Rines was smeared for not allowing *Time* and *National Geographic* to do what they wished with his photographs; Dinsdale was asked to keep his film secret until the Establishment was ready to believe in Nessies, but he was given no indication that this might ever happen; and Burton, perhaps in view of his professional standing in biology, felt free to keep data from the hunters while expecting them to make all of their data available to him.[24]

Inevitably the hunters believe that they have a proprietary right to their own results and want to use them to further their cause. They surely must be reluctant to relinquish them to the Establishment that has given every indication that it would not cherish the data or use them to further the search. And there must also be a very difficult personal dilemma for each hunter: should Nessie become a part of science, the role of amateur hunter will lose its raison d'être; the search will be taken out of the hands of the hunters, who will not in return be guaranteed the rewards of permanent recognition and status that accrue to discoverers who are within the mainstream of science. Indeed, the hunters will not be able to participate in the further exploitation of the discovery thus made. It is therefore not surprising that even as they seek to have biologists take their quest seriously, the hunters also must wish to maintain some control of their data. This attitude makes all the more difficult their attempt to obtain the imprimatur of science.[25]

The distinction I have made between amateur hunters and professional scientists reflects accurately, I believe, the popular view. Clearly, the study of animals belongs to the science of biology, and anyone not a biologist is thereby automatically an amateur. The press has consistently treated biologists as the experts to whom questions about Nessie ought to be addressed; and so the press has characterized as amateurs a number of individuals who have high professional status, albeit in other areas than biology.

In this respect the quest well illustrates how the very success of science has created difficulties when we seek to learn about unconventional and unclassified fields of potential knowledge. Within science the difficulties of doing interdisciplinary work are increasingly well recognized. The very success of research spawns subdisciplines and further specialization; the average scientist concentrates very narrowly, journals restrict their fields of interest, teaching and research in universities are delimited by preexisting definitions and traditions of what each discipline entails, and the view of what legitimate inquiry is tends to be determined by a summing of existing activities rather than by a vision of all that is waiting to be studied. Connections between disciplines are made slowly and always encounter difficulties. Biometry had to overcome the contemporary wisdom in biology that mathematics had nothing to contribute to it;[26] the early biochemists had to overcome the disdain of chemists for the messiness of the systems they studied and also the skepticism of biologists that biologically useful information could come from an understanding of chemical structures; and similarly with sociobiology, psychobiology, and so forth. When a new field of research that falls between recognized disciplines is opened up it always meets with skepticism and resistance—and rightly so, because it has yet to be proved that anything of value will come of it.

The quest for Nessie clearly suffers from this general difficulty and from the inevitable corollary that it is never clear what type of background, credentials, or qualifications best fit a person for this type of work. It is understandable, then, but nevertheless ironic, that some of the Nessie hunters who are denigrated for being amateurs may well possess precisely those qualifications that are in fact best suited to the work at its present stage. The task is to obtain the most definite possible information in the absence of a living or dead specimen and in the absence of any assured way of observing the creatures. Hence sonar searches and attempts to find the best circumstances for serendipitous photography, as well as pressing to the limit the technology for underwater strobe photography—just what the AAS is doing, and Dinsdale, and the LN&MP, and what the LNI did. So too aerial reconnaissance, radar and infrared searches at night, submarines, monitoring with hydrophones, looking for substances that might attract the creatures, designing instruments to get tissue samples if ever a Nessie comes close enough.

Few if any of these activities call for the sort of expertise that biologists command. But in contemporary society we automatically label people according to their formal credentials rather than their talents. We readily forget that science itself was built on the efforts of amateurs and that in observational work amateurs can still make notable contributions: in the discovery of new comets, the exploration of new lands, and the study of animals in their natural habitats. Whether there be justification for it or not, the search for Nessie is tainted in the public view by the nature of the credentials that the monster hunters do not possess.

The quest is difficult, not only because it calls for interdisciplinary effort and unconventional expertise, but because the hunters are trying to do something (or several things) for the very first time. That always poses special problems. Rarely do the pioneers achieve complete success, be it in sport or in science or in geographic exploration. By hindsight we can often see how they could have done better, but hardly at the time. So the hunters should be given credit for breaking new ground even when it did not succeed, for the ideas are often brilliant: nocturnal watches with searchlights or infrared devices; aerial searches, baited cameras, sonar-triggered cameras, and so on—any one of those could have done it, it just happened not to.

We readily forget how difficult it is to do anything for the first time or that things known now were not always known. When Rupert Gould thought of Nessie as a single, trapped specimen of sea serpent he did so in part because, in those years, it was thought that Loch Ness had been a lake for millions of years. A couple of decades later, however, Constance Whyte could think, more reasonably it seems to many of us, of a breeding population, because the history of ice ages and sea-level changes had begun to be worked out. No doubt some more changes of view will have to occur before the mysteries of Loch Ness are solved. But here as in other endeavors the first explorers deserve the credit for making later progress possible. In the often and variously quoted words,[27] even pygmies standing on the shoulders of giants can see further than the giants themselves.

GUILT BY ASSOCIATION

Some subjects are so widely regarded as crackpot that any given phenomenon or idea can be discredited simply by associating it with those typically crankish pursuits. Fringe subjects are commonly found guilty by association in this way, whether or not any evidence is adduced for the association. So Nessie has suffered from the very beginning through a linkage with sea serpents, a quite logical and necessary association. Nessie has also been discredited by unsubstantiated smear and gratuitous association, as in the following examples.

"If any reputable scientist comes forth publicly to back Velikovsky—I for one promise to join anybody who wants to stake out real estate on the moon, build a perpetual motion machine, or equip a safari to search for the sidehill wampus, tripodero, Lochness *[sic]* Monster, or the whirling whimpus."[28] An eminent physicist deplored ". . . public interest in the abstract, the occult, in extrasensory experiences and the Loch Ness monster. . . . The lunatic fringe includes some physicists."[29] The Astronomer Royal was confident that others suffer from hallucinations: ". . . when there is a report of something having been seen which is mysterious and outside ordinary experience, other people begin to think that they see the same thing. . . . The reports of the Loch Ness Monster provide an instance."[30] In the December 1978 *Smithsonian,* an article by Kendrick Frazier, entitled "UFOs, horoscopes, Bigfoot, psychics and other nonsense," lumps together also ". . . the Bermuda Triangle . . . ancient astronauts . . . psychokinetic key benders . . . biorhythm . . . conversations with plants . . . the abominable snowman and the Loch Ness monster. . . ." According to *Time* (12 Dec. 1977, p. 100), "America has been saturated in recent years by tales of the paranormal and claims of the pseudo scientists. . . . Uri Geller, the Bermuda Triangle, E.S.P., levitation, Jeanne Dixon, Kirlian photography, the Loch Ness monster, psychic surgery, Immanuel Velikovsky, thinking ivy plants and . . . flying saucers." Cohen classes together astrology, extrasensory perception, reincarnation, flying saucers, pre-Columbian discoverers of America, Immanuel Velikovsky, Loch Ness monsters, and yetis; and Sladek's list includes Atlantis, Bacon's ciphers in Shakespeare's plays, homeopathy, Loch Ness, Nostradamus, perpetual motion, Velikovsky, Wilhelm Reich, and Zen macrobiotics.[31] Similar references are legion.

Books about anomalous subjects often include Nessie, and some of those books are typically crankish in their gullibility or lack of documentation.[32] Fringe magazines include Nessie in the grist for their mills—*INFO, Pursuit, Fate, Fortean Times*—and the connection is given credence when a monster hunter gives an interview to such a publication.[33] Garish newsstand tabloids[34] help to foster the general idea.

I have discussed elsewhere at some length the pundits' penchant for labeling certain matters pseudoscience by assertion rather than by reasoning.[35] Here I must add that Nessiedom does give the critics some opportunities to make superficially plausible a connection with some of the other topics often asserted to be pseudoscience. So the references in Nessiedom to legends and folklore can seem, to the skeptics, confirmation of an analogy with Velikovsky, fundamentalists, scientific creationists, health faddists, Yoga fanatics, and so on. Yet it is quite illogical to conclude that a claimed natural phenomenon does not exist simply because it is referred to in folklore. As Bayanov has pointed out, if such creatures as Bigfoot exist and are occasionally encountered by humans, those encounters would inevitably be incorporated into epics and legends and myths.[36] Indeed, if there were no references to dragons or lake monsters or sea serpents in folklore, one might equally conclude that Nessies and sea serpents do not really exist.

Dinsdale, Rines, and the Ness Information Service are open-minded about the possibility that the Sasquatch (Bigfoot) is real, and that very defensible stance is used as a basis for ridicule.[37] Potboilers are published that reinforce the suggestion that if one takes Nessie seriously, one automatically does Bigfoot also, and vice versa. *Bigfoot News* further cements the relationship by covering events at Loch Ness.[38] When a group of psychics announces that it will "flush out 'Nessie' and other sea monsters by telepathy," the credibility of the quest is tainted. (That such a venture was claimed to be successful has cast some doubt on the photographs [fig. 15] taken by Shiels, one of the psychics.) Holiday, an enthusiastic monster hunter, has sought to connect Nessie with psychic phenomena and UFOs and he is by no means alone in that.[39]

Another connection between the fringe and Nessiedom can be found in Ivan Sanderson, a biologist whose wide-ranging interests drew him far into gullibility. Sanderson's penchant for wishful thinking is shown in his preface to Holiday's book, where

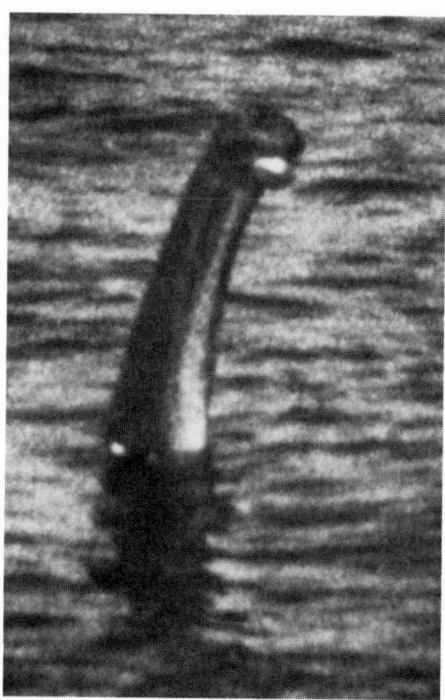

Figure 15. The controversial Shiels photographs. Reproduced courtesy of Fortean Picture Library.

he claimed that science had accepted the reality of Nessies, a statement he repeated later, when he also insisted that an indubitably faked sonar chart shows a marine version of Nessie, some 200 feet long.[40]

There are certain people who are drawn to the strange and out-of-the-ordinary. Some of them do display a high degree of gullibility and a low need for evidence; and some of them number Nessie among their interests. Their involvement in the search, and in discussions of it, discredits the quest, but the hard-headed hunters are powerless to keep their distance from the indubitable cranks and from those who flock to any subject that might serve to shake the foundations of the scientific establishment. Again the vicious circle: since Nessiedom is outside science the crackpots cannot be barred from Nessiedom, as there is no mechanism for doing so; and then, since Nessiedom harbors some cranks, it becomes easy to label the whole enterprise crankish.

HOAXES AND FRAUDS

There is still dispute whether the Piltdown forgery of a "missing link" between apes and humans was a hoax, originally intended only to embarrass certain people when it was publicly revealed, or a full-fledged fraud. In either event, for decades it confused work on the early origins of the human species, wasted the time of many people, and has damaged posthumously the reputations of some eminent people through innuendo and even direct accusation. In Nessiedom an almost endless succession of hoaxes has produced similar damage and has given the conventional wisdom cause to regard the matter as nothing but a hoax, so much so that every new piece of evidence is treated as a possible hoax.[41] As for outright fraud, there is a most visible example in the activities of Frank Searle over a period of some fifteen years.

Newspapers do not like to be hoaxed. When Elizabeth Campbell in 1970 sought references to sightings at Loch Morar, the editor of the *Oban Times* responded to her inquiry thus: "We have consistently refrained from making serious references to the Loch Ness Monster, or any other sub-marine creature supposed to frequent our Highland lochs," giving as reason a hoax perpetrated in 1877. A long memory indeed.[42] The hoaxing of the *Daily Mail* expedition to Loch Ness in December 1933 also did irreparable damage to the view taken by the press of the evidence for Nessie. The *Daily Mail* became the butt of other newspapers' jokes, and the determination not to be taken in (again) can be seen in the tone of jocularity in which events at Loch Ness have been reported since that time. That hoax involved the discovery of a spoor, interpretation of it by the *Daily Mail* expeditioners, and deflation when the tracks were found to have been made by the stuffed foot of a hippopotamus.[43] It is worth noting that the true nature of the tracks was diagnosed, from plaster casts, at the British Museum, which indicates that both the press and the museum were thus prepared, from the beginning, to be very suspicious of anything at or from Loch Ness.

No sooner was the "hippo foot" prank exposed than one of those embarrassing land sightings was reported. A young student of veterinary medicine claimed to have seen, illuminated by the headlight of his motorcycle, a strange creature that crossed the road and splashed into the loch. The student, named Grant, looked over the ground without finding anything; he returned just after daylight, again finding no traces except some flattening of the grass. But soon after his story was made public, some marks of three-toed feet and a pile of bones were found. Apparently someone had thought it too good an opportunity to miss for another joke, hoax, fraud, or whatever. No doubt a similarly undefinable motive caused someone to leave a well-preserved crocodile claw to be found near Urquhart Castle.[44]

During 1933 and 1934, when there was an enormous amount of publicity worldwide about Loch Ness, many other silly hoaxes were attempted, and the tomfoolery has continued up to the present. Within a week of the purported sonar contact in 1954 a faked sonar chart was shown to reporters. News items about the two appeared during the same few days, inevitably confusing almost everybody. In 1961 a self-propelled model of a monster was introduced into Loch Oich, a much smaller loch just south of Loch Ness on the Caledonian Canal. In 1969 the *Daily Mail* was again on the receiving end of a hoax, being presented with a mysterious giant bone found at Loch Ness. The bone, it turned out, came originally from a whale and more recently from a museum in Yorkshire. On April Fool's Day in 1972 a dead elephant seal was launched onto the loch for the benefit of a search party. In 1976 a manufacturer of electronic devices thought it appropriate to display a spurious "sonar recording" from Loch Ness.[45] And so it goes on, damaging the credibility of the quest and the hunters, who can do nothing to prevent it.

Frank Searle took up Nessie hunting in 1969, camping in a tent at Ballacladaich Farm near Dores, accepted by the locals as an honest searcher and as an independent ally by the LNI, who lent him a movie camera. For three years he was unsuccessful, claiming a few sightings but obtaining no photographs. Then his luck changed dramatically. On 27 July 1972 he photographed (fig. 16) two humps low in the water, with what appeared to be the end of a fin or flipper emerging from the water near one of the humps; the dark, warty appearance of the skin fitted what was known of Nessies.[46] Such plausibility could not, however, be granted to the pictures that Searle dates to 21 October 1972: in a sequence of three photos displayed at his tent in 1973, two humps, a neck, and a large head with open mouth remain unmoving while a third hump appears behind and then joins itself to the others (fig. 17). The animal dived, according to Searle, and reappeared on the other side of his boat, giving him the opportunity to "shoot" a head just above the surface, with one and then another hump emerging from the water.

When I was at the loch in the spring of 1973 Searle was not issuing prints of his pictures, telling me a rather muddled story of copyright problems. However, he did not stop me from photographing his display, and later he sold postcards of some of his pictures, as well as posters and pamphlets and an audio cassette of the Loch Ness story as told by Frank Searle.

It is not easy to determine how many photos Searle has

Figure 16. Frank Searle's first photograph purportedly of a Nessie, 27 July 1972. The shape and texture of the object are consistent with Nessie's characteristics. Reproduced with permission of Frank Searle.

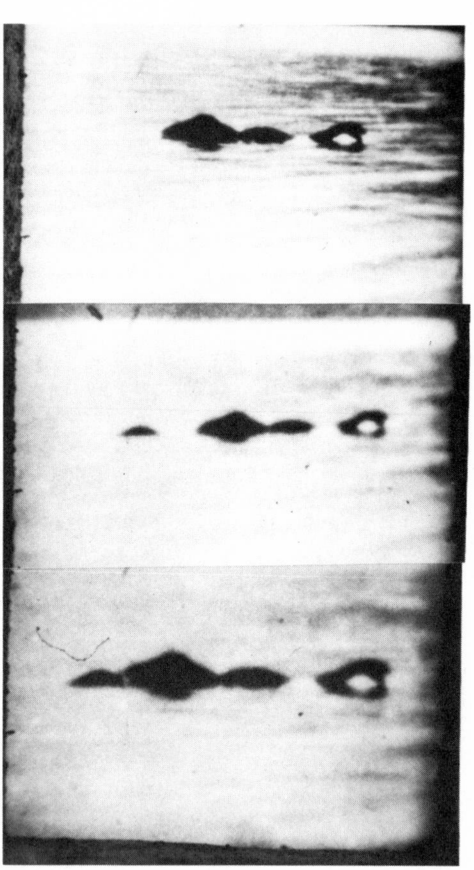

Figure 17. In the spring of 1973 Frank Searle exhibited this sequence of three photographs. Note that the object, with partly open "mouth" and lower jaw just touching the water, appears to remain entirely motionless while a small hump rises in the water and joins up with the rest of the object. The originals were quite sharp; the prints used here were made from Super-8 film, which was the only record Searle allowed me to make at that time. The top photograph has been published in several places (e.g., Searle's *Nessie*); the middle photograph appeared in *Spiegel,* 30 July 1979. Reproduced with permission of Frank Searle.

October 21st 1972 Shot 1
October 21st 1972 Shots 2 and 3

Figure 18. In his book *Nessie* (1976), Frank Searle apparently disavowed the last two shots in figure 17 in a revised description of the sequence of photographs taken on 21 October 1972. Reproduced with permission of Frank Searle.

Figure 19. More of Frank Searle's photographs, reproduced with his permission.

(A) This shot is similar to "Shot 2, 21 October 1972" (see fig. 18) but with an extra little hump.

(B) A sequence of four shots sent to me on 24 November 1973 by Searle.

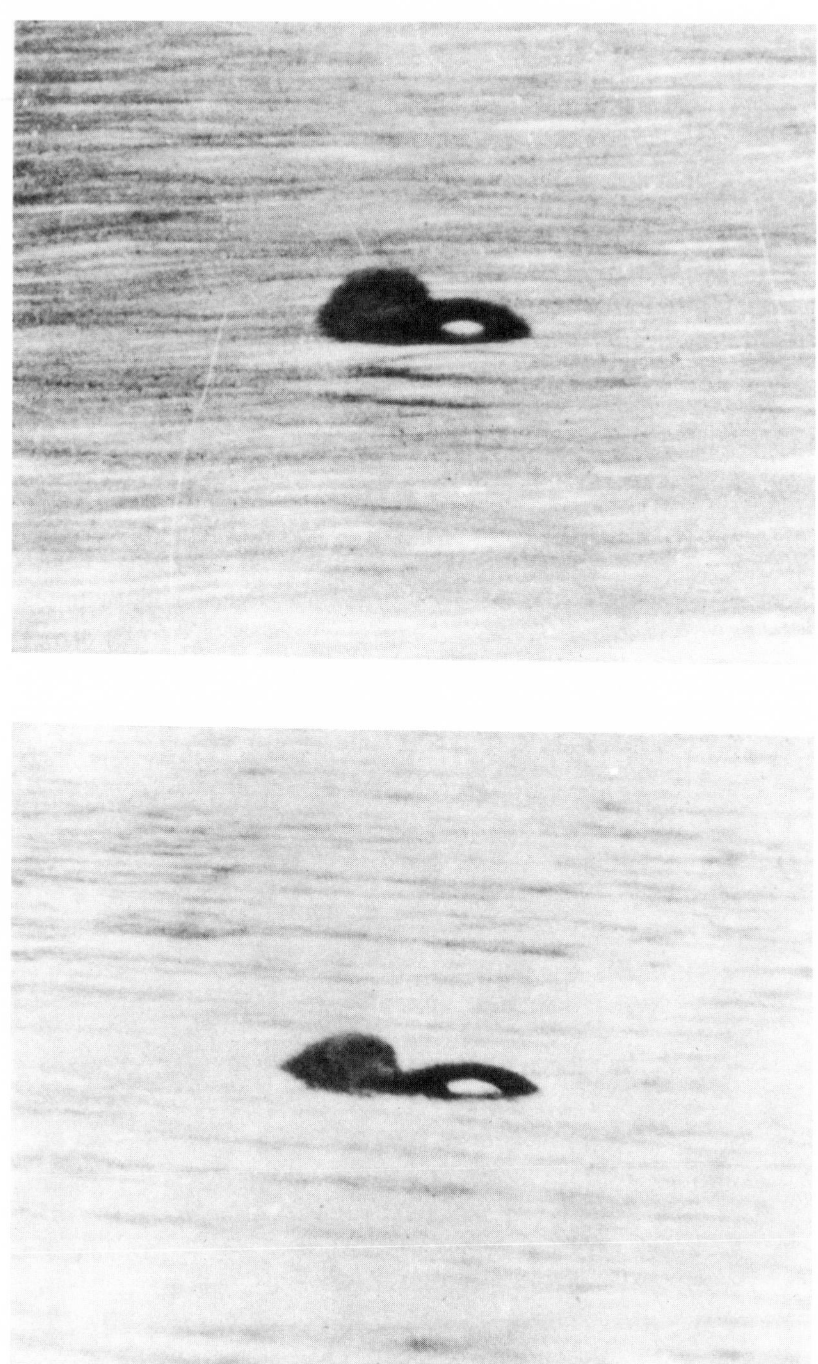

(C) These photographs were published in Searle's *Nessie* (1976) as well as in the *Daily Record* (17 January 1974).

(D) Searle wrote on the back of this photograph, "6.15 AM July 19th 1974 (500mm lens). . . ." When I had it reproduced, the photographer pointed out that the depth of focus far exceeds that obtainable with a telephoto lens. Published in *Frankfurter Allgemeine Zeitung*, 30 October 1976.

(E) A postcard sold by Searle, dated 1975.

October 1st 1975
February 26th 1976

(F) More photographs from *Nessie* (1976).

(G) From Searle's pamphlet *The Story of Loch Ness* (n.p., n.d., copyright 1977, purchased from Searle in 1983), p. 12.

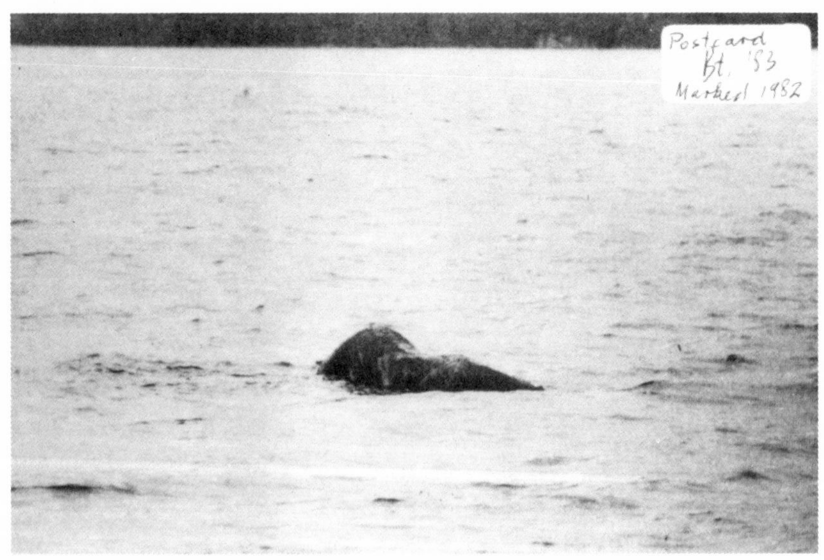

(H) Similar but not identical to (G); a postcard sold by Searle, dated 1982.

produced. In his book *Nessie,* published in 1976, two of the above-mentioned sequence of three have been banished to oblivion, either to before "Shot 1, 21 October 1972" or lost between "Shot 1" and "Shot 2" (fig. 18). But in the spring of 1973 the display at his tent showed them all; and also a neck and head, in color, with snow on the hills in the background; and a long, low hump and a much smaller one a little farther away. Soon afterward, Searle distributed a sequence of four photographs, showing one or two humps changing orientation and with very persuasive splashing of the water. And then there are two pictures showing a little of the body and the neck arching from it and into the water; another head and neck; a hump, head, and neck strikingly similar to those on a postcard (sold locally) of a brontosaurus; and, in 1980, a postcard showing only a powerful tail coming out of the water (fig. 19).[47]

Most knowledgeable Nessie hunters believe that Searle's photographs, at least the great majority of them, are fraudulent, and the embarrassed publishers of his book have refused to keep it in print. Searle's later attempt to publish another book foundered when the prospective publishers followed the good advice of Adrian Shine of the LN&MP. But partly for reasons earlier discussed, Searle's charlatanry long remained unsuspected by many people.[48] Even in 1977 at least two books, Akins's *The Loch Ness Monster* and Cornell's *The Monster of Loch Ness,* showed his pictures without any cautionary warning, as did guidebooks published in 1973 (Robertson's *Loch Ness and the Great Glen*) and 1979 (MacRae's *Handbook—Loch Ness Monster*). A touring guide published in 1977 (Jarrold and Sons' *Loch Ness Country by Car*) mentions (favorably) only Searle in connection with the search for Nessie. And as late as 8 June 1982 the *Sunday Post* (Glasgow) was publishing Searle's photos.

Searle moved from Ballacladaich Farm to a field below Boleskine House, and finally to Lower Foyers, where his caravan bore the official-sounding title "Loch Ness Information Center"; when I saw it, it was full of clippings about Searle and his pictures, with various stenciled commentaries by the man himself that made no mention at all of some of the most important facts about Nessie (for example, Dinsdale's film) and that were scurrilously misleading (about the AAS, for instance). Altogether, for many years visitors to Loch Ness were exposed by the tens of thousands to the unreliable statements of Searle and to his misleading photographs; for those who could not visit the loch, Searle provided a newsletter, sent out quarterly (from Apr. 1977 through Dec. 1983), containing reports of sightings, together with denigrating and vicious comments about Dinsdale, Rines, and others. One can only wonder how much harm has been done to the quest by the likes of Frank Searle.

Notes

1. *Daily Telegraph,* 1 Apr. 1972, p. 15; Anne Robinson and Tony Dawe, "I Dumped Nessie in the Loch," *Times* (London), 2 Apr. 1972; "Loch Ness Yields Another One—12 Feet Long, with Fur," *Times* (London), 1 Apr. 1972, p. 1, col. b.

2. For a recounting of the bombing, see *Scotsman,* 22 Aug. 1983.

3. On the expertise of Rines's team, see John Chesterman and Michael Marten, "Return to Loch Ness," *Omni,* May 1979, pp. 92–95, 123–26; William G. Hyzer, "Zeroing in on Nessie—Part II: New Developments and Approaches Tighten the Imaging Net," *Photomethods,* Oct. 1977, pp. 17, 68.

4. See Andrew Butler, "Dr. Who?" *New Scientist,* 22 July 1982, p. 259; Dennis Meredith, "The Loch Ness Press Mess," *Technology Review,* Mar./Apr. 1976, pp. 10–12; Dennis Meredith, *The Search at Loch Ness,* New York: Quadrangle, 1977, pp. 156–60; "Will Rines Patent Nessie?" *New Scientist,* 4 Dec. 1975, p. 585; Walter Sturdivant, "Loch Ness Monster," *European Community,* Apr./May 1976, pp. 34–40; "Nessie's Return," *Time,* 12 Jan. 1976, pp. 39–40.

5. LN&MP, *Bulletin to Members,* Apr. 1980.

6. Loch Ness and Morar Project, supported and approved by the Scientific Exploration Society. Patrons: Norman Collins; the Hon. Simon Fraser, Master of Lovat; the Rt. Hon. Lord Glendevon, P.C.; David James, M.B.E., D.S.C., M.P.; Sir Robert McEwen, Q.C.; Sir Peter Scott, C.B.E., D.S.C. Field Leader: Adrian J. Shine, F.R.G.S.

7. By 1983 the LN&MP had associated itself with and obtained support from the Drumnadrochit Hotel and its exhibition and was apparently in the process of changing its name to Loch Ness Project. See "Choice—Hunt the Monster," *Guardian,* 14 Feb. 1984; *Loch Ness Project Report* (Official Report of LN&MP), 1983; *Nessletter,* no. 59, Ness Information Service (R. R. Hepple, Huntshieldford, St. John's Chapel, Bishop Auckland, Co. Durham, England DL13 1RQ).

8. On the exorcism of Nessies, see Dan Greenburg, "Japanese Come to Catch Loch Ness Monster," *Oui,* May 1974, pp. 50–52, 82, 112, 114–16, 118–22. The intended use of psychic powers was announced in a press release for 14 Jan. 1977 of the Hoy Organization (Box 57, Paducah, Kentucky).

9. As was the case for the seventy-year-old my wife met in 1980 on a train from Inverness, who had lived all his life near the loch.

10. Elizabeth M. Campbell and David Solomon, *The Search for Morag,* London: Tom Stacey, 1972, p. 181.

11. Mountain reported "no less than seventeen" sightings but also twenty-one photographs during the first two weeks. See *Proceedings of the Linnean Society of London,* pt. 1, 8 Nov. 1934, pp. 7–12; Sir Edward Mountain, "Solving the Mystery of Loch Ness," *Field,* 164 (22 Sept. 1934): 668–69.

12. For the fuss over American commercial interests in the search for Nessie, see David James, "Loch Ness Bureau Control," *Scotsman,* 27 Nov. 1969; David James, "Loch Ness Research," *Scotsman,* 2 Dec. 1969; Anthony Pledger, "Now Transatlantic Breezes Are Ruffling the Enigmatic Waters of Loch Ness," *Scotsman,* 6 Nov. 1969; Anthony Pledger, Reply to Letter by David James, *Scotsman,* 27 Nov. 1969.

13. When the Academy of Applied Science had support from the *New York Times* in 1976, similar strains in the relationship were evident. See Nicholas Witchell, *The Loch Ness Story,* rev. ed., London: Corgi, 1982, pp. 201–2.

14. Loch Ness Phenomena Investigation Bureau, letter to members inviting participants for fieldwork, n.d.

15. LN&MP, *Bulletin to Members,* received Dec. 1979, n.d.

16. Roy P. Mackal, *The Monsters of Loch Ness,* Chicago: Swallow, 1976, pp. 307–8.

17. See Meredith, *Search at Loch Ness,* p. 148; *Nessletter,* no. 17; Witchell, *Loch Ness Story,* 1982, p. 203.

18. For a report of the stonework found in the loch, see Martin Klein and Charles Finkelstein, "Sonar Serendipity in Loch Ness," *Technology Review,* Dec. 1976, pp. 44–57; Chesterman and Marten, "Return to Loch Ness," pp. 92–95, 123–26.

19. See Nicholas Witchell, *The Loch Ness Story,* 2d ed., Lavenham (Suffolk): Terence Dalton, 1976, pp. 217, 221.

20. For a report on the underwater photos, see Robert H. Rines et al., "Search for the Loch Ness Monster," *Technology Review,* Mar./Apr. 1976, pp. 25–40.

21. See Rupert T. Gould, *The Loch Ness Monster and Others,* New York: University Books, 1969, pp. 13–14; Tim Dinsdale, *The Loch Ness Monster,* London: Routledge and Kegan Paul, 1961, pp. 112, 141.

22. On the issue of mainstream publication and public support by biologists, see Mackal, *Monsters of Loch Ness,* pp. 31–32; Virginia Morell, "He Hunts for Living Dinosaurs," *Reader's Digest,* May 1983, pp. 167–72; Witchell, *Loch Ness Story,* 1974, p. 150; Witchell, *Loch Ness Story,* 1976, p. 222; Nicholas Witchell, *Loch Ness Story,* 1975, Harmondsworth: Penguin, p. 105. On the public awareness of the LNI, see Campbell and Solomon, *Search for Morag,* p. 74.

23. See *Proceedings of Linnean Society,* pt. 1, 8 Nov. 1934, pp. 7–12; Witchell, *Loch Ness Story,* 1974, pp. 154–55, 157–58.

24. On the Rines dilemma, see Meredith, "Loch Ness Press Mess," pp. 10–12. See also Meredith, *Search at Loch Ness,* pp. 56–60; "Will Rines Patent Nessie?" p. 585; "Nessie's Return," pp. 39–40. Dinsdale reports keeping his film secret in *Loch Ness Monster,* 1961, p. 112. On Burton's withholding of data, see Peter Costello, *In Search of Lake Monsters,* London: Garnstone, 1974, p. 73; F. W. Holiday, *The Great Orm of Loch Ness,* New York: Avon, 1970, p. 70; Mackal, *Monsters of Loch Ness,* p. 118.

25. The tacit rules in science regarding "ownership" of data and ideas are complex and subtle. Published work can, of course, be used by anyone without asking permission, but proper citation is expected. Matters that one learns of by personal communication, or from reading manuscripts that one is asked to judge for publication or for possible funding, may not be used without express prior permission from the originator. It is regarded as proper to allow discoverers a certain time to develop their ideas before others rush in to use or explain the new data. So, for instance, many scientists looked askance at the manner in which Crick and Watson used results obtained by Rosalind Franklin in Maurice Wilkins's laboratory in their headlong rush to be first with the structure of DNA. See Gunther S. Stent, *Paradoxes of Progress,* San Francisco: W. H. Freeman, 1978, chap. 4: "What They Are Saying about Honest Jim," reprinted from *Quarterly Review of Biology,* 43 (1968): 179–84; James D. Watson, *The Double Helix,* New York: Atheneum, 1968.

26. On the role of mathematics in biology, see Bernard Barber, "Resistance by Scientists to Scientific Discovery," *Science,* 134 (1 Sept. 1961): 596–602.

27. See Robert K. Merton, *On the Shoulders of Giants,* New York: Free Press, 1965.

28. Herbert B. Nichols, "The Velikovsky Excursion," *Christian Science Monitor,* 29 Mar. 1950, p. 18.

29. Samuel A. Goudsmit, "It Might As Well Be Spin," *Physics Today,* June 1976, p. 40.

30. Sir Harold Spencer Jones, "The Flying Saucer Myth," *Spectator,* 15 Dec. 1950, p. 686.

31. Daniel Cohen, *Myths of the Space Age,* New York: Dodd Mead, 1965; John Sladek, *The New Apocrypha,* London: Hart-Davis, MacGibbon, 1973.

32. Charles J. Cazeau and Stuart D. Scott, *Exploring the Unknown,* New York: Plenum, 1979; Cohen, *Myths of Space Age;* William B. Corliss, *Strange Life,* vol. B-1, Glen Arm, Md.: Sourcebook Project, 1976; Sladek, *New Apocrypha.* As exemplifying lack of documentation, see John A. Keel, *Strange Creatures from Time and Space,* Greenwich, Conn.: Fawcett, 1970.

33. For example, see the interview with Roy Mackal in Jerome Clark,

"Tracking the Loch Ness Monsters," *Fate,* (no. 9, 1977): 36–43; (no. 10, 1977): 68–74.

34. For instance, *Secrets of Loch Ness,* no. 1, New York: Histrionic Publishing Co., 1977.

35. See Henry H. Bauer, *Beyond Velikovsky: The History of a Public Controversy,* Urbana: University of Illinois Press, 1984, chap. 8.

36. Dimitri Bayanov, "A Note on Folklore in Hominology," *Cryptozoology,* (no. 1, 1982): 46–48.

37. Peter Byrne, *The Search for Bigfoot,* Washington, D.C.: Acropolis, 1975, foreword, p. 10; "Nessie's Return," pp. 39–40.

38. See Robert Guenette and Frances Guenette, *The Mysterious Monsters,* Los Angeles: Sun Classic, 1975, pp. 17–32; Alan Landsburg, *In Search of Myths and Monsters,* New York: Bantam, 1977; Bauer, *Beyond Velikovsky,* p. 26.

39. For connections between Nessie and psychic claims, see the Hoy Organization press release, 14 Jan. 1977; "Nessie: The Shiels 1977 Photos," *Fortean Times,* (no. 29, Summer 1979): 26–31; Holiday, *Great Orm,* p. 98; F. W. Holiday, *The Dragon and the Disc,* New York: W. W. Norton, 1973; Janet Bord and Colin Bord, *Alien Animals,* Harrisburg, Pa.: Stackpole Books, 1981.

40. Holiday, *Great Orm,* p. xii; Ivan T. Sanderson, *Investigating the Unexplained,* Englewood Cliffs, N.J.: Prentice-Hall, 1978, pp. 5–38.

41. On the Piltdown forgery, see L. Harrison Matthews, "Piltdown Man—The Missing Links," *New Scientist,* 90 (30 Apr.-25 June 1981): 280–82, 376, 450, 515–16, 578–79, 647–48, 710–11, 785, 861–62; 91 (2 July 1981): 26–28. The lack of credence given to new evidence about Nessie is exemplified in Maurice Burton, *The Elusive Monster,* London: Rupert Hart-Davis, 1961, p. 74; Constance Whyte, *More than a Legend,* 3d rev. imp., London: Hamish Hamilton, 1961, p. 9.

42. Campbell and Solomon, *Search for Morag,* p. 102.

43. On the hoax involving the *Daily Mail,* see Gould, *Loch Ness Monster,* 1969, p. 21; Whyte, *More than a Legend,* 1961, pp. 103–7; Witchell, *Loch Ness Story,* 1974, pp. 59–63.

44. The Grant incident is discussed in Gould, *Loch Ness Monster,* 1969, pp. 87–89; Whyte, *More than a Legend,* 1961, pp. 73–76, 107; Witchell, *Loch Ness Story,* 1974, pp. 63, 137. See Dinsdale, *Loch Ness Monster,* 1961, p. 158, for mention of the crocodile claw.

45. The various hoaxes are reported in Whyte, *More than a Legend,* 1961, pp. 9, 18; Witchell, *Loch Ness Story,* 1974, pp. 63, 117, 168, 187; Dinsdale, *The Leviathans,* London: Routledge and Kegan Paul, 1966, p. 60; David James, *LNI Annual Report,* 1969, p. 9; "New Nessie Sonar Tracks Are Fake," *New Scientist,* 12 Feb. 1976, p. 346.

46. On Searle and the LNI, see Witchell, *Loch Ness Story,* 1974, pp. 183–84. His 27 July 1972 photos are in Holiday, *Dragon and Disc;* Barrie Robertson, *Loch Ness and the Great Glen,* n.d. (ca. 1972), 28 pp.

47. See reproductions in James Cornell, *The Monster of Loch Ness,*

New York: Scholastic Book Services, 1977; William Akins, *The Loch Ness Monster,* New York: Signet, 1977; Mary Beith and Ian Sharp, "I'm No Fraud Says Frank the Loch Ness Monster Hunter," *Sunday Mail,* 15 Aug. 1976, pp. 20–21.

48. On the general opinion of Searle among Nessie hunters and his publisher, see Costello, *Lake Monsters,* pp. 112–13; Witchell, *Loch Ness Story,* 1974, pp. 183–84; Beith and Sharp, "I'm No Fraud," pp. 20–21. For another view, see Victor Perera, *The Loch Ness Monster Watchers,* Santa Barbara, Calif.: Capra Press, 1974, pp. 22–31; Lindsey Stringfellow Weilbacher, "More Nessie Sightings," *Rockbridge County Press* (Va.), 15 Sept. 1981, p. 2B; 22 Sept. 1981, p. 1B.

7

The Professional View

The perpetration of hoaxes, the association with fringe subjects, the mixed credentials of the monster hunters—all serve to discredit the quest for the Loch Ness monster in the eyes of the scientific community and the wider society. Even if one were to accept that the best evidence does indeed indicate that Nessies exist, there would nevertheless remain excellent reasons, inherent in the nature of scientific activity, for a lack of participation in the search by biologists.

It is widely believed but nonetheless false that scientists as a class are engaged in a continual search for major new discoveries. The mistaken belief is understandable, since it is a mistake in degree rather than in outright kind. One corrective to this mistaken view might come from recognition that the preponderance of scientific work now being pursued is rather of the sort called "applied" than of the sort called "pure." Although no precise distinction of these is possible, it serves the present purpose well enough to make a distinction in terms of looking for useful things in contrast to simply looking into things that arouse curiosity. Many biologists, for example, are engaged in looking for new knowledge that would be useful in understanding illnesses of various sorts, or in treating them whether or not they are properly understood; others are engaged in finding—designing, even—new strains of useful plants and animals. Most biologists, as most scientists in general, are employed in industry, in hospitals, in independent or governmental research institutes. They work at what their jobs are, not in pursuit of their individual fancies, no matter how inspired or intellectually fascinating or promising. The latter is supposedly engaged in by scientists who also teach at universities, but even there the choice of problems to pursue is determined to a considerable degree by what funds are available—the research must seem of interest to the university's administration or to those who manage the National Science Foundation, the National Institutes of Health, or the like.

That such an interesting potential discovery as Nessie has not brought biologists into the search in droves is not at all surprising since there exist no droves of biologists who are free to drop what they are paid to do and look at whatever happens to interest them.[1] Most scientists work at science in the same manner as most people work at most professions: because they like it, to be sure, but only as a better way to make a living than the available alternatives. Most scientists display what Tullock has so aptly termed "induced curiosity,"[2] that is, curiosity in what they are induced to be curious about by their employers or by grants from foundations or governmental agencies.

There do exist, however, scientists who are primarily driven by their own curiosity, by the simple desire to understand more than is currently understood, and it is people of this ilk who give science as a whole its reputation—because most of the truly famous individual scientists have been such people. But considering even these people reveals that the conventional wisdom has a simplistically misleading view of what is involved in a life of seeking new knowledge through discovery. At any given time some problems are ripe for solution and others are not, because of the existing knowledge in the particular area and in ancillary ones and of the apparatus and techniques available. The ultimately successful scientists rarely seek to leap far into the unknown but rather try to steadily expand what is known, beginning from solid foundations. They look for problems that are solvable—certainly as major and significant as possible, but always in their judgment solvable.[3] As Ziman has nicely put it, science restricts its attention to questions whose answers are capable of being agreed on.[4]

The most lauded discoveries have come to those who discern before others do that a problem is ready to be solved . . . or through sheer serendipity. If a discovery is too far ahead of its time it may have no effect; instead it may lie unused, essentially ignored, until the state of the art has caught up. Such discoveries have been termed "premature" by Stent, and a number of important events seem to fit well into this category.[5] Among them, I suggest, is the purported discovery of Nessie.

Stent's definition of prematurity hinges on the degree to which a new finding can be connected to the existing state of knowledge in a way that permits scientists somehow to use the new finding—to develop a new theory, to carry out a new experiment, to realize the applicability of an old system to a new type of investigation. To be of scientific value a new finding must point the way to potentially fruitful work; it must not be "a statement in a void." The claimed discovery of the so-called Loch Ness monster in the 1930s was for biologists a statement in a void; it did not point to anything they could usefully do. The available information was insufficient to identify the animal, there were no specimens whose morphology or behavior could be studied, and no one could tell how to get such information. And that still applies: reported sightings are far too infrequent, brief, and uninformative to make it reasonable for a scientist to spend at Loch Ness the time needed for the activities of his career. The LNI estimated an average of one sighting for every 350 man-hours of watching; Tim Dinsdale has had but two sightings in more than twenty years since he obtained his film. The sonar and photographic evidence is of no use because it does not permit Nessie to be related to any presently known species. Thus it is not even possible to set up plausible hypotheses about behavior and life cycle that could guide the search by indicating where one might most fruitfully seek to find the animals relatively often. The chance of success through random photography is likewise negligible.[6]

It is therefore natural, inevitable even, that scientists do not join the hunt. I find their apathy, which surprised Dinsdale when he first showed his film, understandable, though monster buffs and monster hunters, from Gould to the present, consider it infuriating, incomprehensible, or reprehensible.[7]

Left to themselves, no doubt the biologists would simply ignore the matter, as they have other premature discoveries—though ignoring the matter in public does not entail an unwillingness in private to entertain the notion that Nessies exist.[8] But they have not been left alone. Instead they are harried by the press for categorical answers, pressed by the hunters with their supposedly incontrovertible evidence, criticized by the fringe buffs for the dogmatism and conservatism of science. The "maybe," "so what," and "where could we go from here?" thus become increasingly defensive and have led to outright rejection. Surely this is understandable in psychological terms: if something is premature it is ignored as though it does not exist; and it is an easy step from acting as though a thing does not exist to believing that it does not. Thus ignoring the premature can easily become resisting it.

Anomalous claims and premature discoveries are discounted also because we cannot imagine how they might fit into the corpus of established knowledge. It is a short psychological step from not seeing how a thing might fit to believing that it cannot fit. Before the nineteenth century it was believed that stones could not fall from the sky, and the empirical evidence was ignored; established knowledge could not imagine by what mechanism stones could come from a place that contained no stones.

The resisting of startling new claims is an inherent part of scientific activity, described with insight and a multitude of examples by Barber two decades ago, in a paper itself ahead of its time.[9] (It was long ignored by sociologists of science and continues to be so by many practicing scientists.) Barber noted that discoveries are resisted when they conflict with established knowledge, particularly when new techniques are involved, and the more so the lower the professional status of the discoverer. On all counts the discovery of Nessie fits that description.

Biologists simply have no notion as to how Nessie might fit into the known classes of animals. The difficulty is well illustrated by Mackal's analysis.[10] He believes that Nessies exist and discusses for eighty pages what they might be in view of what (little) is known of their size, shape, and behavior. He is able to account for between 47 and 88 percent of the known characteristics, but the highest scorer (amphibian) fails on the crucial count of the configuration of Nessie's head and neck, which is rather well established among believers; the second best (eels) fails on that account and also on the equally well established one of the flippers; the third best (plesiosaurs) seems to fail in terms of respiration in water and the low water temperature; and so on. Even a believer who reads Mackal's discussion is disheartened; surely no biologist can be faulted for concluding that Nessies cannot be.[11]

On the matter of techniques, the search for Nessie satisfies Barber's description perfectly. Standard techniques for biological classification involve very detailed study of dead or living specimens; or, on occasion, less detailed observation of species closely related to already well known ones. But in Nessiedom we have only eyewitness reports (almost exclusively from nonscientists), sonar contacts, and photographs of bits and pieces that form much less than a complete jigsaw puzzle. Thus Nessiedom fits again on the last of Barber's criteria, because the discoverers have the lowest possible status in biology—they are altogether outsiders.

Barber was scrupulous to make the important point that the resisting of such discoveries stems inherently from the nature of scientific practice and is not to be ascribed to personal failings of individual scientists. That point is readily lost in the heat of argument, however. It is also typical that those scientists who make the most public statements tend to be less judicious and equable and objective than one could wish. The public is not told truthfully why biologists cannot classify or investigate Nessie; the public is told only that Nessie does not, indeed cannot, exist. So too was the public not told that Velikovsky's science was improbable in the extreme and that he made certain specific demonstrable mistakes, but instead that he was a crank, dead wrong.[12] In a similar vein we do not have scientists differing reasonably on just how safe or unsafe nuclear power might be; instead there are two opposing groups publicly shouting "safe" and "unsafe."

There are reasons why some scientists become excessively dogmatic in such discussions. Having to say, "we simply don't know," is not particularly easy or pleasant. Some may fear (misguidedly, in my view) that an acknowledgment that science doesn't know something might lessen public confidence in science. And some biologists no doubt resent the making of discoveries by outsiders, who do not even stick just to the discovering but who also dabble uninformedly in speculations about taxonomy, evolution, physiology, and so forth. Again, scientists do not care to discuss professional matters in or with the newspapers, particularly not controversial matters. Scientists as a class have a horror of being made to look ridiculous[13]—as can easily happen when one's speculative comments or statements of probability are published in the popular press, always distorted to some extent and shorn of all qualifications and subtleties of nuance.

The public view that science prizes and rewards pure curiosity and disinterested inquiry is overly superficial. Those qualities do characterize much of what is best and lasting in science but only because of the elaborate structure of tacit rules, discipline, hierarchy, and refereeing that determine what eventually is accepted into the corpus of certified scientific knowledge. Status comes to individuals within science for having been right, for the requisite insight, ingenuity, flair, intuition, not for being scrupulously objective and disinterested and intensely curious. So scientists value and guard their reputation for good and sound judgment; and that could easily be lost by venturing into the public arena on the long shot that Nessies are real.

Further, in attempting to connect Nessiedom with existing knowledge the monster hunters and monster buffs make connections that are unwelcome to scientists. A connection with sea serpents and other unidentified creatures, though inherently logical, nonetheless makes scientists uneasy. Connections with old documents, legends, and folklore are perhaps even worse—in some ways, as with scientific creationism, such connections raise the specter that science's victory over religion and superstition is not completely and finally won.[14] And with such connections, scientists know, come those people who are always waiting in the wings to participate in any battle that offers a chance that science might be dealt a stunning blow; those people who rail against arrogant, dogmatic science.[15]

For good and sufficient reasons, then, science does not enter such a quest as that for Nessies, or for sea serpents, or for yetis, until such time as the odds for success are great enough. Individual biologists who do take part do so under difficult circumstances. The venture is thus left largely to outsiders.

MAVERICK SCIENTISTS

The myth is that scientists should and do pursue Truth, in any direction that promises significant new discovery. In reality there is strong, albeit largely tacit, pressure that discourages individual scientists from venturing too far from what their colleagues regard as worthy of study in any given specialty at any given time. Transgressions can be quite costly to one's standing in the profession. So, for example, wise advice to probationary scientists is to choose pretty little *solvable* problems to establish their reputations;[16] and then, if they wish, to become somewhat more venturesome. The long shot, the leap into the unknown or into an entirely new field, is always fraught with danger, even for the scientist with a well-established reputation. So eminent a man as Linus Pauling has suffered professional disdain for going public with his unorthodox opinions about physiology and vitamin C; and unkind things are said within the profession about the long-shot notions of Szent-Gyorgy about cancer.

One of the norms of scientific work, moreover, is "organized skepticism." Scientists are trained to be critical of anything that clashes with reliable, established knowledge. An overwhelming accumulation of indubitable proof is called for to warrant replacing a traditional viewpoint with something new: necessarily so, because a point of view only becomes traditional through having been useful and after surviving many indirect as well as direct tests. That norm of skepticism becomes internalized by scientists, so that they may become individually and by conviction less likely than others to give credence to startlingly novel claims.[17] Indeed, the scientific community may well be the last to give its imprimatur to a revolutionary departure in theory, method, or fact. Biologists thus require more weighty evidence than do nonscientists before they will admit the possibility that such creatures as Nessies might actually exist. And individual biologists take a considerable professional risk by lending public credence to the possibility of Nessies before their colleagues profess themselves ready to do so.

By way of illustration, biologists who venture to mention Loch Ness seriously often emphasize that they express their personal view and not that of their institution; or again, the editor of a scientific journal emphasizes that a particular author's views about Loch Ness are not necessarily those of the journal.[18] Such disclaimers are unusual in science—physicists make no such statement when commenting publicly on matters of physics, nor mathematicians when they speak about mathematics, nor in fact do biologists when they talk of matters biological. Such disavowals are common only when the issues are social or political, or when speakers are commenting outside their field of expertise. That the subject of Loch Ness also stimulates such disclaimers illustrates how strong is the pressure on biologists, from their own intellectual community, not to get involved. A related dilemma has been described by Heuvelmans: If he were to see a Nessie, should he suppress the fact?[19] The pundits would cite his prior belief to impugn his powers of objective observation, yet his observations would be unusually valuable, those of a professionally trained and experienced biologist.

Given that there are risks to biologists in publicly pursuing Nessie, is there any reason why the few interested biologists might not quietly and privately give their help and advice to the hunters? A number have in fact done so. But inevitably such individuals face another dilemma: If they stay totally out of the public eye, and their assistance helps toward the success of the quest, how can they later receive the credit that is their due? This is no trivial matter. In science, credit for ideas is the major currency of which reputations are built, and the practices of signed publication and the scrupulous assignment of credit to others for their ideas and help is strictly adhered to. So the private helper is in something of a quandary. Few are so magnanimous as to desire no public credit in the event of success; for most there must be a continual conflict between disinterestedly helping in a quiet way and wanting to ensure the later acknowledgment of credit, between giving freely and maintaining proprietary control over ideas and data. And that personal conflict, of course, makes relations less than totally comfortable with the amateur monster hunters, who suffer their own, similar personal conflicts. Relations can be strained also because the environment surrounding the hunt lacks the quiet order of the scientific laboratory—see, for example, the case of Dr. McGowan as reported by Michael Enright.[20]

Conflicts and dilemmas of this sort can, I believe, be clearly discerned in the literature of Nessiedom. Maurice Burton's writings become less puzzling if one assumes that he was anxious, in the event that Nessies are real, to be seen as the leading scientific authority, yet also anxious to maintain an escape route should they turn out to be myth or misinterpretation. The less satisfying aspects of Mackal's book (see below) are interpretable in a similar way. One detects an analogous dilemma for the publicly prominent people who help the quest by serving, for example, as patrons of the LNI or the LN&MP: Sir Peter Scott, for instance, helped found the LNI but kept out of the public controversy during the long years of lack of progress, emerging strongly into the limelight when the AAS obtained its first underwater photographs and ensuring permanent credit to himself through suggesting a scientific designation for the Nessies. The one who asked Dinsdale to keep his film secret was perhaps also influenced by the desire to share in the credit should it later become safe to do so.[21]

One way of keeping all options open is to make public statements in a carefully jocular and ambiguous tone, as the *Times* (London) did rather consistently in 1933 and 1934. Then, if the quest founders, it can be pointed out that one had been joking all along; but if it succeeds, an early interest in the matter can likewise be pointed to. So Carl Sagan applied to Nessie the same (rather simple-minded and doubtfully valid) statistical calculations that he had used to attack Velikovsky, in a manner that defies interpretation as to whether he was serious and oblivious to the inapplicability of the model, or joking with poker-faced pen to smear Nessie, or attempting to ensure later credit for his good judgment should Nessie become respectable.[22] Another attempt in a scientific journal to calculate the possible size of the population of Nessies seemed quite straightforward; but when subjected to satire and attack, those authors too sought safety in a certain degree of jocularity about the subject.[23]

The significant general point is that the nature of science makes it so hazardous to give credence publicly to "fringe" subjects that those who do so—be they scientists or not—cannot feel entirely comfortable about it. Therefore they practice great caution; which inevitably permits the sort of carping criticism and questioning of motives that I indulged in above.

In point of fact, it is only wise to be cautious in these matters, for people have lost their positions and their careers for taking seriously, in public, such unorthodoxies. There was passionate, public controversy in 1950 over the ideas ventured by Immanuel Velikovsky; his editor lost his job in the publishing house, and Gordon Atwater was dismissed from the Hayden Planetarium.[24] Denys Tucker, a zoologist, publicly expressed his belief that Nessies exist; he lost his position at the British Museum, and some opinion holds that there was a connection.[25] In private, some who work at that institution have acknowledged that it is decidedly aloof about investigations at Loch Ness and makes clear that members of the staff ought not to become associated with it. One individual there gave private advice to the LNI in the late 1960s and also was invited to join in the work during his vacation: he declined, pointing out that the press would refer to him as being associated with the British Museum and that some undesirable repercussions for him might ensue.

When scientists are cautious in public about claimed anomalies, then, it is not that they freely choose to be mealy mouthed and to mask conviction by jocularity, but rather that they have no professional death wish. Quite often the believer who is also a scientist will adopt something like the following public stance: I do not claim proof of any particular phenomenon or theory—only that a question exists that is amenable to scientific investigation. I do not claim that this is of the greatest interest or priority or that it must be studied—only that it is not unscientific to engage in such study. Some remarkable discoveries in science have come about in unexpected ways; science is and should be open to new suggestions. As for myself, I keep an open mind on this particular issue. I find it interesting, but of course it is an avocation only. Most of my time is occupied by my professional activities, and I dabble only in my spare time in the study of (Loch Ness monsters, extrasensory perception, UFOs, etc.).

So there is a need to be seen as professionally circumspect and decorous. Further, there are internal dissonances for the scientists who have come to take an anomalous claim seriously: one part of them believes, but their professional part finds it difficult to jettison skepticism. That clash might explain the less satisfactory aspects of Mackal's *The Monsters of Loch Ness* (1976); for example, the frustrating analysis that is drawn out and ultimately without a convincing conclusion (pp. 133–217). No doubt the fear of leaping too far led Mackal to place an upper limit of six meters on the length of Nessies,[26] ascribing the many contradictory reports to inaccurate estimates by observers or misidentification of standing waves; he also talked in his book of diameters up to four feet (p. 90) when measurements on Dinsdale's film had already justified five to six feet. One sees how scientists are bound to flounder when they feel obliged to display authoritative insight in the absence of sufficient data; and a desire to ensure personal credit may underlie as well alleged failure to properly acknowledge earlier work.[27] So also the characterization of Mackal's book as much acclaimed and of Mackal as the first scientist to take the problem seriously.[28]

In point of fact, a prominent scientist had publicly taken a serious interest in the Loch Ness monster long before Mackal entered the picture. Maurice Burton of the British Museum, author of many popular books about animals, frequent contributor to the *Illustrated London News,* had come to believe that Loch Ness indeed harbored an unknown type of large animal and had discussed the pros and cons of such possibilities as giant eels and plesiosaurs; his published statements consistently took that stance into the early part of 1960. In June of that year he made a short visit to the loch and thereafter maintained that the Loch Ness monster was largely a matter of misidentification of vegetable mats, otters, birds, and so on—though he kept safe for himself the other option by admitting the possibility of a twenty-foot-long otter-like creature. Burton's public change of heart was dramatic and puzzling enough to have been discussed at some length by Holiday, in an exchange in the *New Scientist,* and elsewhere.[29]

I find it difficult not to interpret Burton's shift as resulting from a desire to be recognized as *the* authority on the subject and to preempt the possibility that Tim Dinsdale might receive acclaim as the discoverer. In 1959 Dinsdale had become intrigued by the problem and had made a lengthy, detailed analysis of eyewitness reports. He also discovered hitherto unnoticed detail in the surgeon's photograph, detail that spoke for authenticity, and he made contact with Burton, who credited him with that newest discovery.[30] It is worth noting, however, that at the same time Burton referred to a critical analysis that he himself had just carried out of the existing eyewitness accounts but failed to mention Dinsdale's analysis. Then in April 1960 Dinsdale achieved the most significant result to date, a moving film of what believers agree is a Nessie hump. He immediately informed the British Museum, where Burton had worked and from which he had recently retired, but then kept his film secret from the public on the advice of an unnamed person.[31] It seems to me likely that it was a person who wanted to find some way of at least sharing in the credit for the discovery. Dinsdale showed his film in private to people with some standing in biology; most likely Burton was one of those people.

Burton mounted a family expedition to Loch Ness in June 1960. He obtained nothing comparable to Dinsdale's film but tried to make much of what little was achieved.[32] By then Dinsdale's film had been shown on television and had aroused great public interest, and Dinsdale began writing his book, *Loch Ness Monster,* which was published in 1961. It seems that Burton was anxious also to prepare for publication a book of his own, *The Elusive Monster*, which appeared a few weeks after Dinsdale's. Indications of Burton's haste can be found in the book itself. For example, some of the classic photographs are represented by sketches and not by actual reproductions; what other reason than urgent haste could there be for this? The same photographs have been reproduced in other books, and they were evidently available to Burton as well; and surely Burton's scientific instincts told him that sketches could in no way substitute for the actual photographs.

How is one charitably to explain Burton's insistence that Dinsdale's film shows only a boat, for which Burton claims nine good reasons that he has unaccountably refused to reveal? (Note that Dinsdale filmed a boat *as a control* after filming the hump; see fig. 5[G] in chap. 2). How charitably to explain the fact that Burton's book gives no inkling that Burton knows Dinsdale? How charitably to account for the speciousness of Burton's semantic analysis of Dinsdale's statements?[33]

Altogether, Burton's actions and his book are sad, easy to interpret uncharitably. The book misleads, misquotes, and misinterprets. For example, Rupert Gould made plain that when he started his investigation in 1933 he expected to find that people had been misidentifying some well-known creature or phenomenon; Burton, however, asserts that Gould went to the loch predisposed to find a sea serpent.[34] Burton ascribes also to Constance Whyte a preformed notion about Nessie even though she had described her dawning conviction as resulting from reports from people she knew and trusted.[35] Burton's book shows not the Gray photo but a sketch of it (fig. 20) that is entirely misleading, representing as the photographed object what is actually its shadow. But perhaps the epitome is the 16mm film made available to Burton by a tourist who thought he had filmed the monster: Burton states that the

Figure 20. A drawing of Maurice Burton's sketch (in *The Elusive Monster* [1961], p. 79), purportedly of Hugh Gray's photograph (fig. 1).

Taylor film looks uncomfortably like an animal and is the most important piece of evidence he has ever examined. Nonetheless he concludes that the film is of an inanimate object, and he has refused to show it to the LNI or to other writers on the subject.[36] Maurice Burton surely exemplifies the dilemmas facing the professional who thinks Loch Ness is worthy of interest.

As the data accumulate it may become less hazardous for biologists openly to declare themselves interested in Nessies. There is some indication that the less rigid structure of American society and American science leaves biologists in the New World freer to follow such interests than are their colleagues in, say, Great Britain. Larry Laudan has suggested to me a more general interpretation: scientists are more aroused to fight the pseudoscience that is close to home than similar ventures elsewhere. So American and British scientists can contemplate with relative equanimity a search for living dinosaurs in Africa, and Americans can be similarly open-minded about Loch Ness, whereas the British find Loch Ness too close to home to be equable over the claim that amateurs have made a remarkable discovery in biology there, in their own backyard. On that reasoning British scientists ought to be more open to the existence of Sasquatch than are the Americans. This interpretation has the major advantage that it can be empirically tested. Laudan points out also that British scientists should not be characterized as generally conservative; they have been very open to taking seriously and to studying claims of psychic phenomena. Yet the cancellation of the Edinburgh symposium and the comments by British scientists reflect a degree of conservatism and unwillingness to deal with outsiders that is not apparent in the remarks made by American biologists about the photographic results achieved by Rines's team.[37]

I find it tempting also to contrast Dinsdale and Rines as representative of two different national styles. Dinsdale's efforts to interest the professionals have encountered the rigidity of British academia, its horror of publicity, its unhurried tempo; and he has responded in the British tradition—where necessary go it alone, with self-discipline and fortitude, a little fatalistic in the recognition that success might not come but nevertheless proceeding in the faith that persistence must ultimately lead to a resolution of the problem. Rines was moved by the American belief that everything is possible and assembled a team of experts with gadgetry galore, wheeling and dealing to enlist individuals with needed talents and facilities, bulldozing his way into at least semiprofessional journals, refusing to let the system and its conservatism determine what happens; with that charming, disarming new-world naivety, he set out to accomplish the virtually impossible. And it is rather nice that these different traditions have come together, that the fascination of the quest has led to a real collaboration, and that the virtually impossible is perhaps in process of accomplishment. If and when a similar resurgence of national traditions makes itself felt in the social and political spheres, the economic and industrial situations of these two nations may also become less depressing to contemplate.

Notes

1. That Nessies are fascinating to so many people does not entail that, even if proven to exist, they would be similarly fascinating to sizable numbers of biologists. Most biologists would continue to find their greatest delight in other things: the study of fruit flies, or snails, or any of the huge number of species that continue to provide challenging questions that await answers—almost always questions whose import is of much

larger significance than the possible existence or nature of one particular species.

2. Gordon Tullock, *The Organization of Inquiry,* Durham, N.C.: Duke University Press, 1966.

3. For a discussion of problem choice and many other aspects of science, see Sir Peter Medawar, *The Art of the Soluble,* London: Methuen, 1967. Reprinted together with *The Hope of Progress* in Peter Medawar, *Pluto's Republic,* Oxford: Oxford University Press, 1982.

4. John Ziman, *Public Knowledge,* Cambridge: Cambridge University Press, 1968, p. 18.

5. Gunther S. Stent, "Prematurity and Uniqueness in Scientific Discovery," *Scientific American,* Dec. 1972, pp. 84–93. See also Henry Bauer, *Beyond Velikovsky: The History of a Public Controversy,* Urbana: University of Illinois Press, pp. 299–301.

6. Tim Dinsdale, personal communication, 1982. William G. Hyzer, "Taking the Odds on Nessie," *Photomethods,* Sept. 1977, pp. 10, 14.

7. Tim Dinsdale, *Loch Ness Monster,* London: Routledge and Kegan Paul, 1961, pp. 8–9, 111; F. W. Holiday, *The Great Orm of Loch Ness,* New York: Avon, 1970, pp. 168–69; Nicholas Witchell, *The Loch Ness Story,* Lavenham (Suffolk): Terence Dalton, 1974, p. 10.

8. For a discussion of biologists' private response to evidence of Nessie, see J. Richard Greenwell and James E. King, "Scientists and Anomalous Phenomena: Preliminary Results of a Survey," *Zetetic Scholar,* (no. 6, July 1980): 17–29; James E. King and J. Richard Greenwell, "Attitudes of Biological Limnologists and Oceanographers Toward Supposed Unknown Animals in Loch Ness," *Cryptozoology,* (1983): 98–102.

9. Bernard Barber, "Resistance by Scientists to Scientific Discovery," *Science,* 134 (1 Sept. 1961): 596–602.

10. Roy P. Mackal, *The Monsters of Loch Ness,* Chicago: Swallow, 1976, pp. 133–217.

11. See Maurice Burton, *The Elusive Monster,* London: Rupert Hart-Davis, 1961, p. 157.

12. For a detailed discussion of the case against Velikovsky, see Bauer, *Beyond Velikovsky.*

13. For example, see Maurice Burton, "The Mystery of Loch Ness," *Illustrated London News,* 8 Dec. 1951, p. 950; Maurice Burton, in *Scottish Sunday Express,* 2 Aug. 1959; Maurice Burton, "What Is the Loch Ness Monster?" *Scottish Field,* Feb. 1960, pp. 48–49; Denys W. Tucker, "Case for the Monster" (review of Dinsdale's *Loch Ness Monster*), *Observer,* 28 May 1961, p. 30; Witchell, *Loch Ness Story,* 1974, p. 198.

14. In Western society, religion discriminated against science well into the nineteenth century. Some individuals and groups (for example, the American Humanist Association) continue to worry about the political clout and social influence of old and new religions, and they see in fringe cults a danger that society is being infiltrated by superstition. To wipe out all traces of the past discrimination and to prevent recurrence of it, these

groups are engaged in what one might call "affirmative action" in defense of reason and objectivity—action inevitably unfair to some and in itself an emotional rather than a rational venture, replete with internal logical inconsistencies.

15. For a discussion of this attitude toward science, see Bauer, *Beyond Velikovsky,* pp. 204–5; Ron Westrum, "Knowledge about Sea-Serpents," in Roy Wallis, ed., *Sociological Review,* monograph no. 27, Mar. 1979, pp. 293–314.

16. See Medawar, *Art of the Soluble.*

17. For example, John Ziman points out that objections to the notion of a positively charged electron were vague, almost unconscious, yet enough to inhibit many excellent scientists from contemplating the possibility. See Ziman, *Public Knowledge,* p. 51.

18. For example, see Robert H. Rines et al., "Search for the Loch Ness Monster," *Technology Review,* Mar./Apr. 1976, pp. 36–37; Roy P. Mackal, "'Sea Serpents' and the Loch Ness Monster," *Oceanology International,* Sept.–Oct. 1967, pp. 38–44.

19. Heuvelmans discusses his dilemma in his book *In the Wake of the Sea Serpents,* New York: Hill and Wang, 1968, p. 31n.

20. McGowan was reportedly distressed at the "Barnum and Bailey" overtones of the expedition with which he was associated; see Michael Enright, "Waiting for Nessie," *MacLean's,* 89 (6 Sept. 1976): 38–46.

21. On Scott's public role, see Dinsdale, *Loch Ness Monster,* 1976, p. 164; "Naming the Loch Ness Monster," *Nature,* 258 (11 Dec. 1975): 466–68. Dinsdale's dilemma is reported in *Loch Ness Monster,* 1961, p. 112.

22. See Carl Sagan, "If There Are Any, Could There Be Many?" *Nature,* 264 (9 Dec. 1976): 497; Bauer, *Beyond Velikovsky,* pp. 224–25. At a symposium at Cornell University in 1976, Sagan reportedly expressed belief that large unidentified organisms exist in Loch Ness. See Kraig Adler, "Loch Ness Monster Evidence Presented at Cornell University," *Herpetological Review,* 7 (June 1976): 41–46.

23. For attempts to calculate the Nessie population, as well as the subsequent ridicule and toning down of the claim, see R. W. Sheldon and S. R. Kerr, "The Population Density of Monsters in Loch Ness," *Limnology and Oceanography,* 17 (1972): 796–97; C. H. Mortimer, "The Loch Ness Monster—Limnology or Paralimnology?" *Limnology and Oceanography,* 18 (1973): 343–45; W. Scheider and P. Wallis, "An Alternate Method of Estimating the Population Density of Monsters in Loch Ness," *Limnology and Oceanography,* 18 (1973): 343; R. W. Sheldon, "The Loch Ness Monster: Reply to Comments of C. H. Mortimer," *Limnology and Oceanography,* 18 (1973): 345–46.

24. See Bauer, *Beyond Velikovsky,* pp. 24, 25.

25. The Tucker controversy is referred to in Heuvelmans, *Sea Serpents,* p. 524; Philip Howard, "A Family of Drop-Outs in the Loch?" *Times*

(London), 15 Mar. 1956, p. 9; Pearson Phillips, "Nessiteras Absurdum," *Observer,* 14 Dec. 1975, p. 11; Witchell, *Loch Ness Story,* 1974, p. 150; Witchell, *Loch Ness Story,* 1976, p. 222; Witchell, *Loch Ness Story,* rev. ed., Harmondsworth: Penguin, 1975, p. 105.

26. This limit set by Mackal is reported in Jerome Clark, "Tracking the Loch Ness Monsters" (an interview with Dr. Roy Mackal), *Fate,* 30, (no. 9, 1977): 36–43; (no. 10, 1977): 68–74.

27. Heuvelmans's review of Mackal, *Monsters of Loch Ness,* in *Skeptical Inquirer,* 2 (no. 1, Fall/Winter 1977): 110–21. Even in science it is not always simple to decide for which statement and to whom credit may be due for earlier work or suggestions; on the fringes of science it becomes very difficult. Regarding Loch Ness there is a major obvious dilemma, one I faced particularly in preparing a bibliography and for which Mackal was open to criticism: Are works dealing with sea serpents and with freshwater monsters in other lakes germane, and should they be cited? If one answers yes, then one has prejudged not only the question of Nessie's existence but also questions of the existence of other lake monsters and of sea serpents, and one even dares to state how closely they are related. If one answers no, then one keeps open those questions but runs the risk of minimizing the important contributions of other cryptozoologists in those related areas.

28. Clark, "Tracking the Loch Ness Monsters"; Mackal, *Monsters of Loch Ness,* dust jacket, p. 22. While authors do not always prevail over the publishers on matters of advertising and jacket copy, they are inevitably held to blame for them.

29. Burton's change of mind is discussed in Holiday, *Great Orm,* pp. 61–71. See also correspondence in *New Scientist:* Peter F. Baker et al., 24 Nov. 1960, pp. 1413–14; 5 Jan. 1961, p. 51; Maurice Burton, 22 Sept. 1960, pp. 773–75; 27 Oct. 1960, pp. 1144–45; 1 Dec. 1960, p. 1478; 8 Dec. 1960, pp. 1549–50; Denys W. Tucker, 27 Oct. 1960, p. 1144; 17 Nov. 1960, pp. 1346–47; Constance Whyte, 27 Oct. 1960, pp. 1145–46; 5 Jan. 1961, p. 51. And see D.W.T., *Observer,* 28 May 1961, p. 30.

30. Dinsdale, *Loch Ness Monster,* 1961, pp. 68–77; Maurice Burton, "The Loch Ness Monster," *Illustrated London News,* 20 Feb. 1960, p. 316.

31. Dinsdale, *Loch Ness Monster,* 1961, p. 112.

32. Maurice Burton, "The Problem of the Loch Ness Monster: A Scientific Investigation," *Illustrated London News,* 16, 23, 30 July 1960, pp. 110–11, 150–52, 192–93.

33. Burton, "Loch Ness Monster: A Reappraisal," pp. 773–75; Burton, "The Loch Ness Saga—A Flurry of Foam and Spray," *New Scientist,* 8 July 1982, pp. 112–13; Burton, *Elusive Monster,* pp. 73–74, 78. See Burton's letters in *New Scientist,* 27 Oct. 1960, pp. 1144–45; 1 Dec. 1960, p. 1478.

34. Compare Rupert T. Gould's *The Loch Ness Monster and Others*, New York: University Books, 1969, and Burton's *Elusive Monster*, p. 14.

35. Again, compare Burton's *Elusive Monster*, p. 15, and Constance Whyte's *More than a Legend*, 3d rev. imp., London: Hamish Hamilton, 1961, pp. xviii–xix.

36. For Burton's reasoning with regard to the Taylor film, see Burton, *Elusive Monster*, pp. 67, 69; Peter Costello, *In Search of Lake Monsters*, London: Garnstone, 1974, p. 73; Holiday, *Great Orm*, p. 70; Mackal, *Monsters of Loch Ness*, p. 118.

37. Compare the remarks of British scientists and American biologists in "Nessiteras Skeptyx," *Nature*, 258 (25 Dec. 1975): 655; letters by Gordon B. Corbet, L. B. Halstead, P. D. Goriup, J. A. Middleton, and Sir Peter Scott, in *Nature*, 15 Jan. 1976, p. 75–76; Phillips, "Nessiteras Absurdum," p. 11; Adler in *Herpetological Review*, pp. 41–46; Rines et al. in *Technology Review*, pp. 25–40.

8

Bad Reasons for Not Believing

In the preceding pages I have acknowledged a goodly number of reasonable grounds for disinterested people to suspend belief in the possible existence of Nessies, even in the face of evidence that now seems compelling to many who have taken the opportunity to obtain and examine it. But the conventional wisdom for these fifty years has also given voice to quite bad reasons for rejecting the possible reality of Nessies.

Most of the commonly voiced, invalid reasons really amount only to the claim that Nessies cannot exist because it is so improbable that they should: various assumptions are made and reasoning from those assumptions is supposed to prove the point. The fallacy of that approach may seem obvious once it is pointed out, but it is nevertheless common in human affairs. Weinberg has cogently illustrated the invalidity of calculating the probability of very improbable events and then calling impossible something that is merely very improbable.[1] In controversies about fringe subjects, the debunkers commonly fall into this trap—in attacking Velikovsky, for instance, the calculations of probability were themselves unsound, and the consequent claims of impossibility (on the grounds of celestial mechanics and Bode's Law) turned out to be incorrect.[2] To prove beyond doubt that things of this sort are impossible is not easy, even if one leaves aside the possibility that miracles might actually happen.

The unsound case against Nessie is nicely illustrated by the criticism that the monster hunters have never asked themselves whether such organisms as Nessies could find support in such an environment as Loch Ness.[3] Of course, as with all rhetorical questions, this criticism is not intended to evoke a response but to make a point indirectly: that since none of the ventured possible explanations—giant eel, surviving relatives of plesiosaurs, large novel mammals—is entirely satisfactory, therefore Nessies do not exist. No useful response is available to be made to people who do not recognize the insupportability of that type of argument. If Nessies exist, explanations of their life-style will be forthcoming in due course; in the meantime, the hunters are merely saying that the various pieces of evidence about the existence of the creatures seem to be incontrovertible. Given animals of unknown type—unknown life cycle, physiology, metabolism, respiration, food, behavior—one simply has no way to determine whether a given environment could support them.

A common variant of that criticism is: if Nessies existed science would have known it by now. This mistaken argument is characteristic of controversies about anomalies.[4] It sharpens the fallacy just alluded to and brings out also the mistaken view quite commonly—though implicitly or unconsciously—held, that all major discoveries have already been made, that science possesses all knowledge in general outline form and only details need to be filled in. At times this misguided criticism is made more particular, revealing a faith that our knowledge of animal species from the fossil record is complete.[5] As a counter, Nessie buffs are fond of pointing out that science rejected the existence of the giant squid until the late nineteenth century, when some freak occurrence caused bits and pieces to be washed ashore in a few places;[6] and we also like the discovery, in the middle of the twentieth century, of the living coelacanth: a creature known by science to be extinct since it disappeared from the fossil record about 70,000,000 years ago but nevertheless still forming, in modern times, a part of the diet of ignorant people on the east coast of Africa.

There is a widespread lack of appreciation of the degree to which such areas as Loch Ness remain *terra incognito*. That people have lived in the vicinity for thousands of years does not mean that science has studied the fauna and flora thoroughly, which in point of fact it has not. For example, the presence of a large population of arctic char in Loch Ness was discovered—beyond doubt, by the way—only by the monster hunters during their investigations.[7] No data on the extent of the migratory salmon population were available, nor on that of the sea trout, nor on that of the permanent population of eels. Even the maximum depth of Loch Ness remains in doubt. Not so long ago a remarkable movement of water was discovered that results from variations in temperature.[8] Clearly, much more remains unknown than is now known.

A related fallacious argument is that there is not enough food in the loch to support a population of Nessies. If these unknown animals exist, no doubt we will at some time discover what they eat and how much of it. For now, reasonable calculations indicate that the supply of fish is great enough to feed a group of large predators.[9]

A specific subtheme of the general fallacy is the (rhetorical) question-criticism, Why are Nessies not more frequently seen? Again, of course, the general answer is that we don't know because we don't know what sort of animal they are; as Constance Whyte pointed out long ago, we do not even understand why they ever come to the surface at all.[10] (A not unreasonable suggestion is that they are predators of fish and surface only incidentally when their prey happen to be near the surface.) Usually this criticism is founded on the assumption that Nessies breathe air because they look like reptiles or mammals that breathe air. In the first place, that assumption may be wrong; in the second place, perhaps Nessies breathe in a very unobtrusive fashion—by means of "snorkel-tubes," or by poking only nostrils out of the water—in which case we may see them quite often but misidentify them as fish rising or as birds fishing. One only believes one has seen a Nessie if something unusually large is seen; very few people have the knowledge of animals to permit them, as Maurice Burton was able to do, to interpret curious sets of ripples as likely indicating a large animal just below the surface, quite possibly moving its head from side to side.[11]

Beyond these answers to the dilemma of infrequent sightings it ought to be recognized that a sighting requires an observer as well as a surfacing. It is perhaps impossible to convey to one who has not attempted to keep a watch at Loch Ness just how daunting a task it is: the loch is a mile or so wide and more than twenty miles long, stretching beyond the horizon from any reasonable vantage point on the ground. Massive objects sometimes cannot be readily seen at a range of a mile unless they contrast sharply with the background. I had the chastening experience, for example, of being totally unable to discern a fishing boat some ten to fifteen feet long as it drifted slowly along the opposite shore, past a rocky cliff with bushes at the waterline; through 7X binoculars, of course, the boat and fisherman were unmistakable. A Nessie would need to show itself in some bulk, or at least at some speed, before one could be sure of noticing it—always provided one was looking in the right direction at the right time.

There is also a popularly held notion that monster hunters line the banks of Loch Ness in never-ending watch, a notion quickly discarded on even a short visit.[12] Certainly there have been organized watches, but only by one or two dozen people at a time—one every mile or so along one side, say—and certainly one comes across hopeful people sitting near the loch, scanning the surface for an hour or so at a time. But when in midsummer of 1980 I spent a week at the loch, I was surprised at how infrequently I saw anyone actually looking at the loch. The weather was rather good for Nessie-watching, too—not very cold and long periods of calm water. I drove several times around the loch and saw a mere handful of people staring at the water. There were many cars speeding along; some picnickers, as often as not with their backs to the water; and in the Clansman Hotel, the only one with a good view of the water from dining room and lounge, the lounge chairs had their backs to the window, and I was the only guest to change the orientation so that I could look out onto the water rather than in toward the piano and the bar.

Nor are there many places along the loch from which a casual visitor even sees the water when the leaves are on the trees. If one drives around the loch the water is in principle visible from some forty-five miles of road, but in practice the trees and shrubbery between road and water mask the view very effectively; one catches glimpses of the water for only about one-tenth of the total distance. Should one wish to stop, there is an average of about one place every mile; all but a few of these vantage points accommodate only one or two cars, however, and most of them have quite a limited range of view of the water. In fewer than two dozen places are there houses or hotels from which one can see the water. It is simply a fact that, despite the continuous traffic, the surface of Loch Ness is not being diligently observed by humans. On the water itself there are a few pleasure boats—sailing, motoring, cruises for tourists—and a few small fishing boats; perhaps as many as a couple of boats per square mile of water at any given time, but in any case the viewing horizon from water-level is not very useful for scanning the loch. Altogether, then, the scene is not that of a popular lakeside tourist resort but that of a sparsely populated area with considerable traffic passing along one not-too-wide, very winding two-lane road.

Finally, the fallacious notion that Nessies ought to be seen more often fails to take into account that the known frequency is not that of sightings but that of *reported* sightings. Westrum, discussing the analogous problem of knowledge about sea serpents, has pointed out that reported and recorded data about such anomalous phenomena reflect not the occurrence of the phenomenon so much as the social demand for reports of that phenomenon.[13] Given the conventional wisdom, witnesses of sea serpents or Nessies are positively discouraged from bearing witness by the incredulity they encounter and the ridicule to which they are exposed. Information about sightings is more likely to be suppressed and lost than recorded and preserved.

In point of fact recorded sightings of Nessies were relatively frequent only during the first flush of publicity in 1933 and 1934, when circumstances were more favorable than at any time since. First, the main road had just been built and there was a clear view of the water for a stretch of about twenty miles before the bushes and trees grew again between the road and the water. Second, the conventional wisdom had not yet hardened into supercilious disbelief; in fact, there was a social demand for reports of sightings. The record of sightings from Loch Morar is also instructive in this connection. After a spectacular event there in 1969, a few people canvassed locally for information and rather quickly collected reports of thirty sightings, twenty of them since 1964, a startlingly large number given that Loch Morar has no road along it and only a handful of houses.[14] Had the Loch Morar Survey not done this investigative work, the uninformed skeptic would still be able to say that the lack of reports from there proves unequivocally that Morag does not exist.

A popular argument with debunkers is the unreliability of eyewitnesses and how easily one can be misled. The general argument, of course, has real substance; but applied to the record at Loch Ness it is much weaker. Dinsdale, Owen, and others have acknowledged the general point, but it ought to be recognized that these are acknowledgments of temporary mistakes that the witnesses themselves quickly corrected.[15]

The various suggestions of common creatures and inanimate objects that might have been mistaken for the monsters are as improbable as the existence of Nessies. The most absurd explanations were fittingly disposed of by Gould, Whyte, and many others since;[16] if Nessiedom were science, all of that would be a part of the standard literature and would not require continual reargument. But as it stands the same insupportable suggestions continue to be made. One class of absurdities includes whales, sharks, seals, and so on, common enough creatures that are easily recognized by most people. We are asked to believe that whenever such creatures show themselves in Loch Ness they are invariably misidentified, so that over a period of fifty years no one has ever recognized them for what they actually are.[17] Otters are less well known to most, but it still seems rather unlikely that even a hypothetical giant otter would be so consistently misidentified; and the long neck and large flippers associated with Nessies are decidedly anomalous for otters.

Maurice Burton's favored mats of vegetation, propelled by trapped gases, have been rather nicely countered by Holiday:[18] how improbable that their random movements should always take them away from the watcher and cause them to submerge, so that no one has ever seen an identifiable mass of rotting vegetation at close range and no such mat has ever been stranded. To a certain extent these hypothetical mats suffer the same improbability as the hypothetical tree trunks: how strange it is that these inanimate objects have the propensity at Loch Ness to take on the shape of an animal. Applied to the underwater photographs of 1975, Witchell has made the point sharply:[19] amid tens of thousands of blank frames of film are a few showing fish and a half-dozen showing some peculiar objects, one of them very much like a head (see fig. 9 in chap. 2), another like a long neck attached to a body with two visible appendages (see fig. 8 in chap. 2). How strange that the only waterlogged trees carried by hypothetical currents past the cameras on these several separate occasions should so well mimic not just an animal but even one that looks like the usual description of a Nessie.

Another popular shibboleth assumes implicitly that Nessies are a fraud and ascribes the continuance of the myth to the commercial instincts of the tourist industry.[20] One cannot deny that the souvenir shops sell models of Nessies in cloth, clay, and plastic, and Nessie T-shirts, emblems, and guidebooks. But all that is small potatoes; a determined effort to make money from the presence of Nessies would surely involve hotels and campgrounds overlooking the loch, viewing platforms with coin-operated binoculars, boat marinas everywhere—all the things that are simply not to be found around Loch Ness.

The local planning committee restricts most effectively the construction of commercial facilities at the loch. The Clansman Hotel has about twenty bedrooms with a view of the loch (through small windows set rather high in the walls) and a dining room and lounge with a fine view of the water (as mentioned before, with the lounge chairs facing away from the loch). The small hotel at Foyers has a nice though distant view; and the tiny private hotel called Tigh-na-Bruich has a few windows from which to look out on the water, as does the youth hostel at Alltsigh. There is also a campground not far from the southern end of the loch, and a hotel set far from the water claims good views there. Now that is hardly the way to cash in on the presence of a tourist attraction gambolling in the water. I have already mentioned the paucity of parking spots with views: note that it took until 1980 for a small parking lot to be built above Urquhart Castle, supposedly one of the best points for Nessie watching. Also in 1980 permission was given to build the first new jetty on the loch for many decades, again at Urquhart Castle, to serve the Jacobite boats that offer two or three cruises per day between Inverness and Urquhart Bay. I can only conclude that if Nessie is employed by the tourist industry, then it is a bumbling industry indeed, allowing or forcing the LNI to vacate its site when it was attracting more than 50,000 visitors during the season.

The last type of fallacy I want to mention is based on a lack of appreciation of the physical difficulties of the quest. This fallacy asks "why don't they" and rhetorically refutes the existence of

Nessies because there are so many "easy" ways to find them. Why not catch one in a net? Because trawling is not permitted in the loch and permission would not be given for this particular attempt, even if someone were to underwrite the huge expense involved. Moreover, it is not so clear that one could successfully net creatures capable of traveling at speeds of ten miles per hour or more, in water deeper than 700 feet in many places, and with a total area to cover of about twenty-five square miles.

Why not try sonar? That has been done, successfully on numerous occasions,[21] but the information is of no help in identifying the creatures; it only reveals that large objects are moving about. Moreover, the capabilities of sonar should not be overestimated: if the Swedish Navy, with the finest up-to-date equipment, cannot stay in sonar contact with a foreign submarine cornered in a fjord,[22] then amateurs with borrowed equipment might not always be able to find Nessies in the loch. (Incidentally, I enjoy suggesting a different possible explanation for the Swedish episode: it was not a submarine at all, it was one of Nessie's marine cousins which became invisible on sonar when its air-supply became low!)

Why not hydrophones? That also has been done; a few unexplained sounds were obtained on tape, but what to do with them? One hydrophone may even have struck a Nessie as it was lowered into the water,[23] but that furthers the quest not one iota. Why not submarines? They have in fact been used.[24] A sonar contact or two were made, but the animals are much faster than the minisubs available to the hunters. (The U.S. Navy has not yet offered the services of one of its machines, which might in fact have a fighting chance.) Visibility is very low on account of the peat-stained water, with about thirty feet apparently being the present limit with available light sources.

Why not infrared observation at night? Or aerial reconnaissance in the daytime? Or extraction of tissue samples with some sort of dart gun when finally a Nessie comes within range? Why not bait Nessies with ground-up fish, blood, or sex attractants? All of those things have been tried,[25] within the constraints imposed by lack of continuing financial support. Virtually all of the work at Loch Ness has been done by amateurs, with equipment and supplies borrowed or rented or purchased with the hunters' personal resources. Why not underwater photography? Precisely what the

Academy of Applied Science has been doing since 1972, against staggering odds and with remarkable successes in 1972 and 1975. Yet the difficulty of this approach is illustrated, for example, by the utter failure of the photographic experts from the National Geographic Society.[26]

Quite simply, no one has yet thought up a practicable and certain method for locating and observing Nessies, which says more, however, about human abilities than about the existence of Nessies. Perhaps some other explanation of happenings at Loch Ness will turn up than the presence of Nessies; if so, it will not be because Nessies cannot exist, or would find the water too cold, or should be seen more often, or do not have enough to eat, or would float to shore when dead—or for any other reason that is drawn not from nature but from human theorizing about nature.

There are good, simple, overriding reasons for not believing that Nessies exist: direct, incontrovertible proof of their existence has not been established, and the existence of creatures of this sort is highly improbable in light of our current reliable knowledge of biology. The bad reasons mentioned above add nothing to this case. In point of fact those bad reasons are not why people disbelieve, they are merely specifications of some of the improbabilities. The disbelievers choose not to entertain the possibility that this particular improbability might turn out to be so; and they adduce the aforementioned bad reasons when arguing against the believers and seeking to persuade others to their own viewpoint. Such arguing cannot be productive, cannot lead to resolution of the argument, because it misses the crucial points.

One important disagreement is over the issue of whether one may or ought to believe something that is improbable. I have some comments about that in the following chapter; here I shall restrict myself to making the obvious point that it is not always unproductive or foolish to believe something that is improbable: some of the greatest achievements were improbable before the event, and surely most of us have benefitted in our personal lives from things that were improbable in the extreme before they happened.

The disagreement, then, reduces to the question of whether one may or ought to believe this particular improbability. I cannot understand why not. No harm seems to have come to people in consequence of believing that Nessies exist; at the very least, the disbelievers have not shown that such harm naturally follows from belief. Ultimately, it seems to me, the disbelievers misguidedly invoke those bad reasons because they wish everyone else to share their own estimates of probability and their own worldview. By contrast, I think it desirable that all reasonable viewpoints should flourish in areas where no clear certainty has been established.

Notes

1. Alvin M. Weinberg, "Science and Trans-Science," *Minerva,* 10 (Apr. 1972): 209–22.

2. Henry H. Bauer, *Beyond Velikovsky: The History of a Public Controversy,* Urbana: University of Illinois Press, 1984, pp. 69–71, 313–14.

3. This criticism is made in Roger Tippett, "Loch Ness: Un monstre mal nourri," *La Recherche,* 7 (no. 65, Mar. 1976): 278–81.

4. Ron Westrum, "Science and Social Intelligence about Anomalies: The Case of Meteorites," *Social Studies of Science,* 8 (1978): 461–93.

5. See Constance Whyte, *More than a Legend,* 3d rev. imp., London: Hamish Hamilton, 1961, p. 178.

6. See Bernard Heuvelmans's discussion of the giant squid in his book *In the Wake of the Sea Serpents,* New York: Hill and Wang, 1968, chap. 2.

7. See Peter F. Baker and Mark Westwood, "Under-Water Detective Work," *Scotsman,* 14 Sept. 1960.

8. This so-called seiche in Loch Ness is reported in *Scotsman,* 14 Oct. 1960; see also Tim Dinsdale, *Project Water Horse,* London: Routledge and Kegan Paul, 1975, pp. 149–50.

9. R. W. Sheldon and S. R. Kerr, "The Population Density of Monsters in Loch Ness," *Limnology and Oceanography,* 17 (1972): 796–97.

10. Whyte, *More than a Legend,* 1961, p. 179. Several mistakes are combined in the argument that if Nessies were plesiosaurs they would be frequently seen at the surface and we would find bodies of dead ones cast ashore (see Roy Chapman Andrews, *All about Dinosaurs,* New York: Random House, 1953, p. 128; L. B. Halstead et. al., "The Loch Ness Monster," *Nature,* 15 Jan. 1976, pp. 75–76). First, since our fossil record of plesiosaurs ends about 65,000,000 years ago, who knows how they might have evolved in the meantime? Second, it seems unwise to be too certain about the life-style of creatures that we know only from fossilized skeletal remains. Third, some plesiosaurs seem to have shared with, for example, crocodiles the practice of swallowing rocks, which would prevent carcasses from floating (see Angus d'A. Bellairs, *Reptiles,* London: Hutchinson's University Library, 1957, pp. 92–93; Michael A. Taylor, "Plesiosaurs—Rigging and Ballasting," *Nature,* 290 [23 Apr. 1981]: 628–29). Fourth, there are not many shallows and beaches around the loch to accommodate such flotsam: along much of the perimeter the

shore slopes steeply into the water; and a few walks along the beaches reveal a dearth of dead fish washed ashore, though there is no doubt that the loch is well stocked with trout, char, sea trout, salmon, and eels.

11. See Maurice Burton, "The Problem of the Loch Ness Monster: A Scientific Investigation," *Illustrated London News,* 16, 23, 30 July 1960, pp. 110–11, 150–52, 192–93.

12. The fallacy of a never-ending watch is mentioned also in Daniel Cohen, *A Modern Look at Monsters,* New York: Dodd, Mead, 1970, p. 99.

13. Ron Westrum, "Knowledge about Sea-Serpents," in Roy Wallis, ed., *Sociological Review,* monograph no. 27, Mar. 1979, pp. 293–314.

14. A report of the canvassing for sightings at Loch Morar is in Elizabeth Montgomery Campbell and David Solomon, *The Search for Morag,* London: Tom Stacey, 1972.

15. Tim Dinsdale, *The Loch Ness Monster,* London: Routledge and Kegan Paul, 1961, pp. 81–82, 110; William Owen, *Scotland's Loch Ness Monster,* Norwich: Jarrold and Sons, 1980. See also chap. 1.

16. Rupert T. Gould, *The Loch Ness Monster and Others,* New York: University Books, 1969, chap. 2; Whyte, *More than a Legend,* 1961, pp. 183–85. See also David James, "The Monster Again," *Field,* 14 June 1962, p. 1160; Philip A. Stalker, "Nessie—A Comparison of Fact and Fiction," *Weekly Scotsman,* 30 Mar. 1957, p. 5.

17. In December 1984 a seal did make its way into Loch Ness, and it was subsequently seen by many people. That it was recognized immediately as a seal underscores the point, as does the fact that this was the first time in living memory that a seal had been sighted further up the River Ness than the vicinity of Inverness (*Highland News,* 12 Dec. 1984; 24 Jan. 1985; 2, 16 Feb. 1985; 9 Mar. 1985; 4 Apr. 1985; 9 May 1985). But the event was barely mentioned in other Highland newspapers (*Inverness Courier,* 8 Feb. 1985; *Aberdeen Press and Journal,* 14 Feb. 1985) and not at all in other British papers. I am indebted to J. W. (Dick) MacKintosh for these clippings. The lack of coverage by national papers illustrates the difficulty of getting information about the Loch Ness story if one is not at the scene or in touch with local residents.

18. F. W. Holiday, *The Great Orm of Loch Ness,* New York: Avon, 1970, p. 65.

19. Nicholas Witchell, *The Loch Ness Story,* 2d ed., Lavenham (Suffolk): Terrence Dalton, 1976, p. 220.

20. Curtiss MacDougall makes that assertion bluntly in *Hoaxes,* New York: Macmillan, 1940, pp. 13–14.

21. The use of sonar in the search for Nessie is reported in Dinsdale, *Loch Ness Monster,* 1982, pp. 178–84; Adrian Shine, "The Biology of Loch Ness," *New Scientist,* 17 Feb. 1983, pp. 462–67.

22. The Swedish incident is reported in William J. Broad, "Strategic Lessons from an Elusive Sub," *Science,* 218 (29 Oct. 1982): 450–51.

23. Dinsdale, *Project Water Horse,* pp. 143–44. See also Roy P.

Mackal, *The Monsters of Loch Ness,* Chicago: Swallow, 1976, pp. 55–56, 61–64, 76–78.

24. Dinsdale, *Project Water Horse,* pp. 81–90.

25. See Dinsdale, *Project Water Horse;* Mackal, *Monsters of Loch Ness.*

26. William S. Ellis, "Loch Ness—The Lake and the Legend," *National Geographic,* June 1977, pp. 758–79.

9

Bad Reasons for Believing

It is not only the disbelievers, of course, who wish to persuade others to their own view; the believers also wish to do that, and they too have or invoke some bad reasons. Among those may be a desire to see the authority of science broken, to replace the scientific establishment with something else, to have science be something other than what it happens to be. Such motives are commonly discernible in arguments on the fringes of science—for instance, in the Velikovsky affair—but I find them in the Loch Ness controversy only in rather weak form, as in the ambition that science admit eyewitness testimony.[1] For the rest, one does see some resentment against certain scientists for their specific comments and some general disappointment that science has not engaged in the hunt. But no great general hostility toward science is evident.

The literature offers many examples of people who have come to believe that Nessies exist because they themselves claim to have seen one. Now that is a common but bad reason for believing all sorts of things. I am one of that large number of people who have seen what were thought to be sea serpents or UFOs; but I am also one of that smaller number who had the opportunity to discover that they were mistaken. There are surely many who have been similarly misled but had no opportunity to discover that. Personal experiences subjectively interpreted have led people to believe many things that are doubtfully true (extrasensory perception, say), some things that are doubtful in the extreme (ghosts, psychokinesis, astrology, for instance), and some things that are outright frauds or outright wrong (some physical and spiritualist mediums, for example). It is also not valid to say that personal experiences are satisfactory reasons for believing provided other people have had confirming personal experiences of their own—that holds for all the examples just mentioned. My own belief in Nessies is grounded on film and photos and the pattern of confirmatory sonar and eyewitness testimony and so forth, but I have never (so far as I know) seen one.[2] No doubt a personal sighting would remove all shadow of doubt for me, but it ought not to (unless it were at closer quarters and with more of Nessie showing than in any sighting of which I have heard).

I suggested earlier that a belief in the reality of Nessies is not harmful. Some militant debunkers of claims of anomalies maintain that our society and culture are at some risk if occult and superstitious beliefs become widespread. With that I would not disagree if it is a question of quackery in medical matters (psychic surgery, extreme forms of faith healing, laetrile, and so forth) or of cults led by fanatics or impostors. On many other matters I doubt that more harm ensues from belief in anomalies (Bermuda Triangle, pyramidology, or UFOs, say) than from the degrees of ignorance and gullibility and poor education that influence so much of our political and economic and social behavior. Belief in cryptozoological phenomena—Loch Ness monsters, sea serpents, Bigfoot, Mokele-Mbembe (dinosaurs in Africa)—seems to me singularly harmless.

The significant question is whether one is led through belief in one anomaly to believe in an increasing number of others. The literature gives some indication that such a progression can come about, though I doubt it is frequent let alone inevitable. Some tendency in that direction is admittedly inherent in the pursuit of anomalous phenomena. We do strive to make connections among our beliefs, and we do seek validation of our beliefs by reference to already established beliefs. When science or the conventional wisdom do not offer a connection we sometimes try to validate one anomaly by connecting it with others. So believers in the Bermuda Triangle suggest "explanations" in terms of UFOs or extraterrestrial influences or postulated "energy fields"; and Kirlian photography gets connected with "bioenergetic fields" or "bioplasma" or "morphogenetic fields" or auras or other claimed psychic phenomena. Even in cryptozoology there are a few who suggest connections with UFOs or with psychic phenomena—some have "explained" Nessie and Bigfoot and even more improbable entities (the Black Dogs of England, Mothman, and the like) as creatures from another dimension, as effects of "space-warps" or "time-warps," as manifestations of some sort of "energy" at some sort of "frequency." But it is not inevitable that such connections be made. Those who are fascinated by the possible reality of a particular anomaly may be quite capable of critical thought and may be quite skeptical of other anomalies: in the preface to *Fads and Fallacies in the Name of Science,* Martin Gardner told of the correspondents who agreed with him that most anomalies are spurious even though individuals had a favorite to which they gave credence.

The risk of cutting off one's own beliefs more and more from established ones may be illustrated in extreme form by the life of Wilhelm Reich, who proceeded from unorthodoxy in psychoanalysis and in communist ideology to spurious discoveries in biology, to the even more spurious discovery of orgone energy and the deployment of that energy in battles with UFOs.[3] But such an extreme case makes no general argument against unorthodoxy or eclecticism; rather, it makes the argument for critical and analytic thought. I differ with the militantly global debunkers since it seems to me that they do not want others to be analytic and critical, they merely want others to believe as they do about acupuncture, parapsychology, UFOs, and the rest.

I think that my fascination over a quarter of a century with the question of Nessie has been accompanied by, even if it may not have brought about, an increase in my ability to analyze and to discriminate; and I have attained to some degree the ability to hold some beliefs firmly and to act on them while remaining aware that the beliefs may be mistaken. Nessie has done me much more good than harm, and I cannot imagine why that cannot be so for others.

Nevertheless, there are bad reasons for believing and some possible dangers to unqualified belief in matters not only not established but improbable to boot. In contemplating the possibility of improbable things it can be helpful to discriminate different types as well as levels of improbability and to be particularly on guard when, for example, claimed anomalous events are "explained" by anomalous theories. One does well to recognize that the onus of proof lies on the claimants of anomalies; that extraordinary claims also call for extraordinary proof. In these connections Marcello Truzzi has offered useful discussion and schemes for classifying anomalies.[4] *Cryptoscientific* claims refer to anomalous facts—unicorns or Nessies, say—which do not fit our expectations but which would not call for a revolutionary change of our theories if they were to become established. *Parascientific* claims refer to anomalous processes or relationships and suggest modes of explanation different from accepted ones. Immediately one then recognizes the "doubly anomalous" nature of crypto-parascientific claims, in which anomalous events are supposed to be "explained" by anomalous theories. Such combinations I have mentioned above as a dangerous tendency to relate one anomalous claim with another, and referred to earlier (chap. 6) as one factor that makes more difficult Nessiedom's battle for respectability—as when Holiday sought to relate Nessies to UFOs and to psychic phenomena, or when Shiels promised to make Nessie appear through psychic means. But those are anomalous minorities within cryptozoology and within Nessiedom. The quest for Nessie can be carried on intelligently and without rejecting the corpus of accepted, reliable knowledge in the various scientific disciplines.

Perhaps the bad reasons for believing in Nessie, or in any other improbable anomaly, all stem from one bad reason: a wish to believe. (I have detailed elsewhere how the arguing over anomalous claims provides fertile ground for wishful thinking.[5]) A wish by definition is a subjective entity, a desire for emotional satisfaction, the existence of which by no means entails that reality offers any possibility of satisfaction. Many of us have wished, for example, that some person should come to love us; when that does not happen, we had best accept the fact and not proceed to act as though reality had conformed to our wish. So in pursuing the possible reality of an anomaly we must be on guard that our thinking and observing and arguing remain analytic and not become wishful. That calls for incessant vigilance, for it is characteristically human to lapse into "thinking wishly."[6]

Notes

1. On Velikovsky, see Henry H. Bauer, *Beyond Velikovsky: The History of a Public Controversy,* Urbana: University of Illinois Press, 1984, pp. 60–61, 204–5. On the use of eyewitness testimony, see William B. Corliss, *Mysteries Beneath the Sea,* New York: Thomas Y. Crowell, 1970, p. 139; Tim Dinsdale, *Project Water Horse,* London: Routledge and Kegan Paul, 1975, pp. 183–84; Nicholas Witchell, *The Loch Ness Story,* Lavenham (Suffolk): Terence Dalton, 1974, p. 79.

2. In August 1983 my wife (through binoculars) and I (through a telescopic lens) saw a black shape appear and disappear several times in deep water in Urquhart Bay, and waves seemed to break over the object. Neither of us thought it was animate. But looking at the Super-8 film I shot through the 70mm telephoto lens, we both realized that it could have been animate. Or it was a very unusual wave-effect, like the succession of "humps" in the Tait photograph (see fig. 10 in chap. 5), except that there was only *one,* which appeared and disappeared several times. Either it was a Nessie or we have evidence for the appearance of something in Loch Ness that could certainly be so interpreted and whose real identity is anything but obvious.

In the spring of 1985 I glimpsed for a few seconds the head of the seal that had been in the loch for several months. Or was it the seal?

3. A good summary of Reich's case is given by Peter Marin, "What Was Reich's Secret?" *Psychology Today,* 16 (Sept. 1982): 56–65. A full-length study by Colin Wilson, *The Quest for William Reich,* Garden City, N.Y.: Anchor/Doubleday, 1981, is more balanced than most and contains a good bibliography.

4. See Marcello Truzzi, "Definitions and Dimensions of the Occult: Towards a Sociological Perspective," *Journal of Popular Culture,* 5 (no. 3, Spring 1972): 635–46; Marcello Truzzi, "Parameters of the Paranormal," *Zetetic,* 1 (no. 2, Spring/Summer 1977): 4–8; Marcello Truzzi, "On the Extraordinary: An Attempt at Clarification," *Zetetic Scholar,* 1 (no. 1, 1978): 11–19; Marcello Truzzi, "On the Reception of Unconventional Scientific Claims," in Seymour H. Mauskopf, ed., *The Reception of Unconventional Science,* Boulder, Colo.: Westview Press, 1979.

5. The relationship between anomalous claims and wishful thinking is discussed in Bauer, *Beyond Velikovsky,* pp. 195–203.

6. David Starr Jordan, *The Higher Foolishness,* Indianapolis: Bobbs-Merrill, 1927, pp. 14, 61.

10

Nessie, Science, and Truth

My aim has been to show that there is nothing inexplicable about the continuing controversy over Loch Ness: the disagreements and the ways in which they have been pursued and all else having to do with the enigma are straightforward consequences of human nature and of the nature of human institutions as they grapple to know and understand the world. Through the differing interpretations in chapters 1 and 2 it should be clear that a given fact will be assigned quite different significance depending on what one is looking for, or expects to find, or is predisposed to believe. That is in the nature of things, especially when evidence is circumstantial. And it is also in the nature of these things that the identifying and collecting and verifying of the evidence are all problematic.

Most fundamentally, however, the enigma of Loch Ness illustrates the dilemma of contemporary society when confronted with knowledge that is not science. We have become so enchanted with the special power and reliability of scientific knowledge that we now demand the imprimatur of science before we are prepared to believe anything. So when "scientific" tests—not just any tests—demonstrate the efficacy of fluoride-impregnated toothpaste, we believe them. And when Maurice Burton spends a few days at Loch Ness with his family, he can dignify and certify his findings through the happy choice of a title for his articles: "The Problem of Loch Ness: A Scientific Investigation." Not just *an* investigation, but a *scientific* one. Note, by the way, that the only possible warrant for that description is that Burton himself would have been described by most people as a scientist; but such a title as "A Scientist Investigates" already carries rather less rhetorical force—let alone, as would have been even more accurate, "A Retired Scientist Investigates."[1]

Although we often forget the distinction, science is not synonymous with truth, nor is truth synonymously science. Many things scientific have been untrue—all superseded theories, for example, and all present ones that await becoming superseded. And there are important truths that are certainly not scientific: love between people, for instance. In the quest for Nessie it often seems as though the hunters seek the imprimatur of science to certify that Nessies exist, yet that imprimatur would add nought to the evidence, let alone to the truth of the matter. The scientific establishment, refusing to concern itself with Nessiedom, protests that Nessies cannot, or do not, or have not been proven to exist, as though there were a necessary connection between the proven existence of Nessies and the studying of Nessies by science.[2] So, on that general issue, hunters and protagonists share a common viewpoint with scientific bureaucrats and with disbelievers—the view, wrong and indeed insupportable, that science and truth are invariably and closely linked, if not identical.

The precise nature of what we call science has been much discussed, yet it can hardly be said that agreement has been reached among scientists, historians, philosophers, sociologists, and others about what "ought" to be called science, about what criteria (if any) distinguish science from other intellectual activities. If one leaves aside these attempts to *prescribe* what science is, there remain a few useful *descriptive* accounts of what is, in common usage, classed as science.[3] Notably, though scientific knowledge has no special claim to being absolutely true, it is marked by an uncommon degree of reliability—follow the rules and you can predict eclipses, cure diseases, transform heredity, perform innumerable acts that, not so long ago, would have been taken as clear proof of miracle working.

That distinction between truth and reliability is not at all paradoxical; it simply stresses that scientific activity is marked by scrupulous consistency and the unremitting attempt not to take things for granted. Although the assumptions and hypotheses used have no claim to absolute truth—whatever absolute truth might in point of fact mean—the data and results and conclusions are reliable *within the frame of those assumptions and hypotheses.* Science is reliable because it does not stray far from what has already been established and accepted as known.

Scientists deal with three kinds of things: with events or phenomena or facts or measurements or observations; with methods or techniques, by which those facts are established; and with regularities or laws or hypotheses or theories, by which those facts are "explained"—fitted together. Common scientific practice is marked by the keeping of two of those three sorts of things firmly in the realm of the already known, seeking novelty only in the third aspect. So, for example, the analytical chemist will accept the contemporary state of the art in chemical theory, and in chemical methodology, and can thus with precision and reliability determine what substances and how much of them are contained in a sample; or, seeking to develop more sensitive and precise methodology, he will again accept the theory and this time the facts—working with samples of known content, trying new methods and regarding them as valid if they yield the right results. To take a more specific instance: around the turn of the century application of long-proven methods had established how much radiation of various frequencies (or wavelengths) is emitted by a "black body" at various temperatures. Starting from there the entirely new theory of quanta was developed to explain those reliable facts established by reliable methods.

By contrast, matters that are commonly called anomalous phenomena or pseudoscience (depending on one's preconceptions) are characterized by a lack of firm grounding in the known facts *or* methods *or* theories; indeed, all three of these features are doubtful to a considerable degree, or are claimed to be novel, or are even counter to expectation. Take modern-day astrology, for example. There are no accepted facts that demonstrate relationships between time of birth and personality or life history; nor does the drawing up of horoscopes follow any method that is widely agreed (even among astrologers) to be valid; nor is there a plausible theory that would explain astrological correlations.

Speculative work in science occasionally involves a lack of well-established grounding in two (but not all) of these three characteristics: the search for gravitational waves, for instance, involves looking for a new type of fact by methods quite unproven, but the theory of gravitation and of gravitational waves is accepted in science.[4] That particular search has so far been a failure, which is most often the case with these speculative ventures. They are called, in science, "high risk" because the danger is considerable that valuable scientific time will be wasted and that the investigators' reputations will suffer (or, at the very least, not improve). Linus Pauling's attempts to establish the efficacy of vitamin C, as well as a relationship between nutrition and mental health, are contemporary instances. His methods are the accepted ones of biochemistry and clinical chemistry, but his theories are eclectic and his claimed facts are disputed.

From this point of view one sees how the search for Nessie is unlike science. Certainly the facts are in dispute—whether Nessies exist, and the status of the evidence about that. Theory in biology also does not recognize the possible existence of such creatures in this era and in the claimed environment (see, again, Mackal's attempted analysis[5]). And the methods being used are either unacceptable in science generally or inadequate or unacceptable in biology—that a new species exists cannot be established by eyewitness testimony or tantalizingly vague photography or transitory blips on a sonar screen, though sonar and photography are well established as reliable in other areas of science (even of biology). So, considering the three relevant characteristics (facts, methods, theories), Nessiedom is not firmly grounded in two and only doubtfully grounded in the third—speculative or novel in two-and-a-half, let us say. Nessiedom is thus more speculative even than most "high-risk" ventures in science, though less unfounded perhaps than such a typical pseudoscience as astrology. All these, of course, are merely other words for the earlier discussions of why Nessiedom is not science.

But do Nessies actually exist?

As I said earlier (preface and chap. 3), that is not the question I have sought to argue in this book. My hope has been to illuminate why the matter has been controversial, and why the controversy has persisted, and why science has remained disengaged—and, at least implicitly, what must happen if science is to become engaged. It is my hope too that I have indicated that one can usefully employ the same type of critical analysis when other such issues engage one's interest: the possible existence of other undiscovered creatures (a field now called cryptozoology[6]) or of such other claimed phenomena as UFOs or ESP or acupuncture or Kirlian photography.

Although science has no answers at the moment, that does not entail that no answers exist, for science is not truth, nor is truth science. One would do well to realize this, and to be aware also that one can reach a satisfactory personal belief about such matters. I shall conclude, then, by trying to justify as rational, albeit not scientific, my personal belief that Nessies exist.

I came across Dinsdale's *Loch Ness Monster* in the early 1960s and thereafter read whatever else I could about the matter, without reaching a firm conviction for something like a decade. Dinsdale's film could not be dismissed, it seemed to me, particularly after the examination of it by the Intelligence Centre of the Royal Air Force.[7] Yet there was somehow not enough known for me to be certain, and I had a general awareness of the possibility of being wrong about matters of this sort—for example, if Piltdown Man could be a hoax. . . .

In the spring of 1973 I met Tim Dinsdale at Loch Ness and he gave me his article about the "flipper" photograph.[8] I arranged for him to speak in 1975 at universities in and around Kentucky and thus came to know him and saw his film many times—including the computer-enhanced sequence that clearly shows a second hump (fig. 6, in chap. 2). Since then I have had no serious doubt that Nessies exist, for Dinsdale happens to be as reliable a person as I ever hope to meet. Yet I am clear that this state of belief can be satisfactory only to me. It is a *personal* belief and not one that meets the required standards of proof in science. But that does not trouble me, because I comfortably believe many things that are not known to science and find those beliefs useful and reliable in my life, satisfactory bases for action and for further thought.

Given my belief, I find it easy to see confirmations everywhere: in continuing successes with sonar contacts, in the work of the Academy of Applied Science, in Heuvelmans's *In the Wake of the Sea Serpents* (1968). As time passes I find that my belief depends less and less on the manner in which I came by it—a general phenomenon, for individuals and for science. In Nessiedom some pieces of evidence that helped to capture belief have later been set aside or even rejected without altering the belief: so Dinsdale, before his first expedition, was greatly influenced by the surgeon's photograph, among others, yet nowadays he can contemplate the possibility that those photographs may not be valid without thereby ceasing to believe that Nessies are real.[9] I am indebted to Larry Laudan for making clear to me the generality of this phenomenon in science—to give just one example, that we accept Newton's laws (within, of course, their limits of applicability) for very different reasons now than the ones that led to the initial acceptance of those laws.

I can accommodate both my trust in Dinsdale's evidence and my belief that Nessies exist to my belief that Lester Smith (see chap. 1) was truthful in reporting that he and his associates invented the monster and "arranged" for it to be sighted. Genuine sightings over centuries have been very rare, a handful or several handfuls only in a given year. Until 1933 sightings aroused little public interest; thus the report of 1930, published in the *Northern Chronicle,* a regional newspaper, elicited only a little published correspondence.[10] As Westrum has pointed out on several occasions,[11] reports of anomalies require not only the occurrence of anomalous phenomena but also that the witnesses feel some encouragement to testify: the media must be receptive if the wider public is to hear about the evidence. Once a report is published, earlier witnesses feel encouraged to relate publicly their earlier experiences; and if the media so decide, then a full-fledged public flap and controversy ensues (as illustrated in the matter of UFOs[12]).

Lester Smith and his colleagues in 1933 found a way to attract wide public attention to the claims of a monster in Loch Ness; were it not for their efforts the sightings might have continued to be of passing local interest only. But whatever spurious report Smith may have created, the result in my view was that it stimulated the recounting of *genuine* earlier and contemporary sightings and a determined effort, by many more people than ever before, to keep a watch. Quite in general anomalous phenomena are often what Westrum has called "hidden events"; and even a fraud or a hoax or a mistaken observation can then lead to genuine information becoming public. That a fraud may have brought about the public flap of 1933 does not entail that Nessie does not exist. Indeed, it may have required such an artificial stimulus to generate sufficient interest that people would be prepared to spend time in the pursuit of such an elusive phenomenon. Perhaps, then, I ought to have included the name of Lester Smith[13] among those of the pioneers to whom I dedicate this book.

Clearly there is no formula or algorithm by which I can show that the evidence for the existence of Nessie is sufficient. Those who wish to reach conviction, one way or the other, must pursue their own train of thought and reach their own estimate of probabilities. The important lessons, I think, are these: one should maintain some awareness that one's own beliefs, though rational, may turn out to be wrong; and one ought to remember that the conflicting beliefs of others are not necessarily false, and that even if they are false they may nonetheless be quite rational.

Notes

1. Maurice Burton, *The Elusive Monster,* London: Rupert Hart-Davis, 1961, p. 15.
2. Science has often engaged in searches for things that may not exist—in particle physics, for instance, it happens all the time.
3. For descriptive accounts of what science is, see, for example, Henry H. Bauer, *Beyond Velikovsky: The History of a Public Controversy,* Urbana: University of Illinois Press, 1984, chap. 15; Michael Polanyi, *Knowing and Being,* Chicago: University of Chicago Press, 1969; John Ziman, *The Force of Knowledge,* Cambridge: Cambridge University Press, 1976; John Ziman, *Reliable Knowledge,* Cambridge: Cambridge University Press, 1979; John Ziman, *Teaching and Learning about Science and Society,* Cambridge: Cambridge University Press, 1980.
4. See H. M. Collins, "The Seven Sexes: A Study in the Sociology of a Phenomenon, or the Replication of Experiments in Physics," *Sociology,* 9 (1975): 205–24; H. M. Collins, "Son of Seven Sexes: The Social Destruction of a Physical Phenomenon," *Social Studies of Science,* 11 (1981): 33–62.
5. Roy P. Mackal, *The Monsters of Loch Ness,* Chicago: Swallow, 1976, pp. 133–217.
6. For a definition of cryptozoology, see Peter Costello, *The Magic Zoo,* London: Sphere, 1979, pp. 17–24. For details of cryptozoological work, see *Cryptozoology* (journal of the International Society of Cryptozoology, P.O. Box 43070, Tucson, Az. 85733).
7. See the appendix in Peter Costello, *In Search of Lake Monsters,* London: Garnstone, 1974.

8. Tim Dinsdale, "The Rines/Edgerton Picture," *Photographic Journal,* Apr. 1973, pp. 162–65.

9. Tim Dinsdale discusses the possibility in *The Loch Ness Monster,* 4th ed., London: Routledge and Kegan Paul, 1982, pp. 199–206.

10. See the letter from "Piscator," *Inverness Courier,* 29 Aug. 1930; letters from "Camper," "Not an Angler," "Invernessian," *Northern Chronicle,* 3 Sept. 1930; letters from "R.A.M.," "Another Angler," *Northern Chronicle,* 10 Sept. 1930.

11. For example, see Ron Westrum, "Social Intelligence about Anomalies: The Case of UFOs," *Social Studies of Science,* 7 (1977): 271–302; Ron Westrum, "Science and Social Intelligence about Anomalies: The Case of Meteorites," *Social Studies of Science,* 8 (1978): 461–93; Ron Westrum, "Knowledge about Sea-Serpents," in Roy Wallis, ed., *Sociological Review,* monograph no. 27, Mar. 1979, pp. 293–314; Ron Westrum, "Social Intelligence about Hidden Events," *Knowledge: Creation, Diffusion Utilization,* 3 (Mar. 1982): 381–400.

12. See Westrum's discussion of UFOs (note 11).

13. Steuart Campbell, who saw an earlier draft of this book, pointed to my lack of consistency here. I have commented that Dinsdale and Whyte, for example, by protecting the identity of witnesses, would inevitably carry less conviction with their readers, yet I have done the same thing in referring to "Lester Smith." The latter's death has now, I believe, made it legitimate for me to be more explicit. The quotation in chapter 1 is from *Marise* (London: Peter Davies, 1950, p. 95), by Stephen Lister. In corresponding with me, Lister revealed that his name was actually D. G. Gerahty. He was an acclaimed novelist, not only as Stephen Lister, but also as Robert Standish.

APPENDIX A

A Brief Chronology of the Loch Ness Story

BEFORE 1933

Water-spirits are common in folklore and mythology. In Scotland, water-kelpies, water-bulls, and water-horses have traditionally been associated with lochs and streams; kelpies were particularly associated with streams, water-bulls and water-horses with lakes.[1] Two questions must be answered: Does the folklore have any basis in real animals? Are there stories about creatures specifically in Loch Ness?

Certainly some people believed water-kelpies and water-bulls to be real creatures; indeed, attempts were made to capture them.[2] There are a few references published before 1933 to such creatures specifically in Loch Ness.[3] Almost all the books and many of the articles about Nessie refer to Adamnan's biography of St. Columba, who encountered a water-monster in the River Ness. A floating island on Loch Ness could have been in actuality the large humps of Nessies at the surface, and there is even an early mention of Morag of Loch Morar, also likened to an island.[4] But I can produce nothing written before 1933 that unequivocally refers to large, nonmythical animals in Loch Ness, other than the items of 1930 in the *Northern Chronicle* and the *Inverness Courier*.

During and after the publicity of 1933 and 1934 there were published second- and thirdhand accounts of sightings during the nineteenth century and early in this century. But there were also statements from local and other knowledgeable people that there had *not* been any tradition of strange creatures at Loch Ness.[5]

1933 AND 1934

Beginning in May 1933 Highland papers carried reports of large animals seen in Loch Ness. In October the national papers picked up the story, and by the end of the year the publicity was worldwide. Tourists and correspondents visited the loch in large numbers. Locals as well as visitors and far-flung pundits divided into believers and disbelievers. Joking about the monster was commonplace, alcohol and the tourist industry being frequently referred to.

The contemporary newspaper accounts give quite a variety of descriptions of the purported "monster," but when Rupert Gould interviewed most of the witnesses he found their descriptions reasonably consistent; he also reported that the newspapers had characteristically distorted and sensationalized the accounts.[6] The consensual description was of humps like an upturned boat, movement with or without splashing, and occasionally a longish neck and a small head. Sightings were almost always on warm, calm days. Sir Edward Mountain organized a watch by a couple of dozen people for several weeks in the summer of 1934; more than twenty sightings were reported by the watchers.

Apart from a few truly bizarre suggestions, almost everyone thought in terms of one animal that had chanced into the loch from the sea during exceptionally high water in the River Ness. References to sea serpents and plesiosaurs were soon made by the believers; the disbelievers offered as explanation birds, eels, otters, and seals, and such even more unlikely candidates as crocodiles, sharks, squid, and whales, as well as inanimate things—logs or masses of vegetation. There were a few purported sightings on land. And there were some hoaxes—tracks made with the preserved foot of a hippopotamus, collections of bones left at the water's edge.

Some photographs were published, including Hugh Gray's (fig. 1 in chap. 2) in the *Daily Record* in December 1933, a still from Malcolm Irvine's film in January 1934, and the surgeon's photo (fig. 2 in chap. 2) in April 1934; those three were also published in

Gould's book of 1934. Photographs taken by Sir Edward Mountain's watchers were reproduced in the *Illustrated London News* (18 Aug. 1934) as well as in newspapers, and something (fig. 13 in chap. 5) photographed by some anonymous person appeared in the *Illustrated London News* (1 Sept. 1934) as well as in various newspapers. A motion picture shot by Mountain's team was shown to the Linnean Society of London, the only consensus among the viewers being that something was moving in the water—otter, seal, whale, and "none of the above" all had their champions.

1935 TO 1949

The international and national publicity quickly declined after 1934. Highland newspapers published only a few reports of sightings, a half-dozen or so per year. The matter was by no means forgotten, however. Loch Ness was mentioned in many discussions of water monsters or sea serpents, and several scientists and science writers left open the possible reality of Nessie.

Another film was obtained in 1936 and shown as part of a Scottish newsreel; a still appeared in the *Sunday Mail* on 1 November 1936 and in the *Daily Mail* the following day. In 1937 there were reports of two or three monsters being seen at the same time, as well as some baby monsters. Tourists were sufficiently interested that the pamphlet published by the Abbey at Fort Augustus was revised and expanded in 1938.

In 1941 the Italians claimed to have killed the Loch Ness monster during an air raid. In 1942 *Time* published a report that a dead basking shark, twenty-four feet long, had been found on the shore of Loch Ness. In 1947 several prominent citizens of Inverness testified to recent sightings, and it was estimated that 1,000 people had seen Nessie since 1933.

FROM 1950 TO 1959

Serious, widespread interest in the possible reality of Nessies was revived by Constance Whyte, who published first in a privately circulated magazine, then in the *King's College Hospital Gazette* (1950), then in a pamphlet, and finally in the book *More than a Legend* (1957). That book is still worth reading for its judiciousness and common sense and attention to all aspects of the matter—natural history, folklore and myth, the nature of scientific activity and the role of the amateur, the evidence about Nessies. Whyte's work marked a change from thinking in terms of a single extraordinary creature, strayed from the sea or monstrously misshapen, to contemplating the existence of a breeding population landlocked after the last ice age. *More than a Legend* also made widely available the second surgeon's photo (fig. 2 in chap. 2) of 1934, the Stuart photo (fig. 3 in chap. 2) of 1951, and the Macnab photo (fig. 4 in chap. 2) of 1955. More than 2,300 copies of the book were sold in Britain in 1957 and more than 400 abroad, and sales continued at an average rate of nearly 200 a year for at least the next eleven years.[7] Although this did not make it a best-seller, it is clear enough from the later literature that Whyte's work was crucially significant in stimulating the determined and partly successful later efforts at Loch Ness.

During the 1950s Maurice Burton admitted the remote possibility that descendants of plesiosaurs had survived in Loch Ness, though he also suggested giant eels as candidate monsters. Burton's colleague at the British Museum of Natural History, Denys Tucker, publicly expressed belief in the reality of Nessies in 1959 and stimulated the Oxford and Cambridge expedition of 1960. More than one group of divers attempted a search. A one-man kayak expedition at night produced a classically disputable photo—the Cockrell photo (fig. 12 in chap. 5); another search threatened to attempt to kill a specimen and later produced another classically disputed photo—the O'Connor photo (fig. 11 in chap. 5).[8] The BBC even tried (unsuccessfully) to enlist the Navy's expertise in a search for Nessie.

British television held a "trial" of Nessie in 1951. In 1957, during a BBC broadcast from the loch, a sonar echo was recorded strongly enough to suggest a large creature. The first generally acknowledged echo from a Nessie had been obtained by chance by a fishing trawler in 1954; within a few days another trawler generated a hoaxed echo. Boy Scouts floated a wood-and-canvas "monster" on the loch in 1958 and fooled a number of people. Also in 1958 it was revealed that a large clawed foot had been found floating in the loch—back in 1937! The foot had every appearance of being from a large crocodile.

In 1950 it was suggested that sightings of Nessies might actually be of mines laid in the loch in 1918. In 1956 the British Museum found shadows on the water to be responsible for Nessie sightings. In 1958 the Italian journalist Gasparini claimed to have invented Nessie (supposedly toward the beginning of Aug. 1933; by that time, however, as the records show, others had already anticipated Gasparini).

In the mid-1950s the Abbey at Fort Augustus was logging 30,000 visits a year by tourists, and the Abbey's booklet about the monster was in its fourth edition. Although serious interest in the matter had been revived, eyewitnesses continued to be ridiculed. Bishop Hoffmeyer recounted the patent amusement of the Edinburgh bookseller who sold him a copy of Whyte's book.

1960 TO 1976

Constance Whyte's writings had stimulated a flurry of activity at Loch Ness. In April 1960 Tim Dinsdale shot what remains the best moving film of a Nessie; the showing of that film on television in June greatly increased public interest. Students from Oxford and Cambridge mounted expeditions in 1960 and 1962. In the latter year the *Observer* supported a yacht-based expedition by Hasler. Also in 1962 the Loch Ness Phenomena Investigation Bureau (later Loch Ness Investigation, or LNI) was founded by Norman Collins, Richard Fitter, David James, Peter Scott, and Constance Whyte. For about ten years (up to 1972) LNI organized teams for surface photography during each summer, enlisted the cooperation of others with special expertise, and tried many things: aerial surveillance by glider and autogyro, night observation by searchlight and infrared, sonar from shore- and water-based units, attraction of Nessies through baiting, underwater searches by submarine.

In the mid- to late 1960s Roy Mackal helped to obtain considerable support, notably from Field Enterprises (publishers of *World Book Encyclopedia*). Robert Rines and the Academy of Applied Science joined the hunt in 1970, and their work continues to the present. The LNI ceased operations in the early 1970s; its files have been deposited with the Loch Ness and Morar Project. In 1974 Rip Hepple of the LNI founded the Ness Information Service, whose bimonthly *Nessletter* has been since that time the only regular source of reasonably reliable information about doings at Loch Ness.

Reports that Loch Morar harbors Nessie-like creatures were dramatically underscored in 1969 when a boat supposedly collided with Morag. The Loch Morar Survey was organized, gathered eyewitness accounts, and carried on watches in 1970, 1971, and 1972. Adrian Shine, who later organized the Loch Morar Expedition, made underwater searches of the loch for signs of large creatures, to depths of between 30 and 100 feet; and underwater surveillance with television cameras was tried, visibility under water in Morar being very much better than in Ness.

The scope and vigor of all this activity was not matched by results. LNI obtained several pieces of suggestive and interesting, but not definitive, films, yet the strongest evidence was still Dinsdale's film, which David James was able to have analyzed by the aerial photography experts of the Royal Air Force. Sonar contacts were made on several occasions by different teams using different instruments. The best new photographic results came from the underwater cameras of the AAS in 1972 and 1975.

Several important books and many articles and newspaper accounts appeared in the 1960s and 1970s, and the tone of much of this writing was for the first time more accepting of the existence of Nessies than dismissive.[9] Important in this connection also was the publication of Heuvelmans's book, *In the Wake of Sea Serpents,* the English edition of 1968 being widely and respectfully reviewed. During the 1960s the Loch Ness monster was one of the ten most popular subjects of inquiries directed to *Encyclopedia Britannica* (the other nine, according to a UPI story on 29 Dec. 1969, being sex education, gun control, pollution, drugs, space achievements, student unrest, food from the oceans, the Arab-Israeli crisis, and the Great Depression).

Nessie's public respectability declined again after 1976. David James, Robert Rines, and Sir Peter Scott had scheduled a symposium of scientists for Edinburgh in December 1975, to discuss the evidence and in particular the AAS photos of 1975. But newspapers revealed information about the photos before the event, some of the participants withdrew, and the symposium was cancelled (instead there was a presentation of the evidence to invited scientists and journalists in the Grand Committee Room at the Houses of Parliament in Westminster and a publication in *Nature*). The *New York Times* supported the hunt in 1976 and gave much favorable publicity to it during that year, but definitive results were not obtained and the *Times* returned to its previous stance of ignoring Loch Ness or joking about it.

Although much was accomplished in the quest between 1960 and 1976 there were also developments less pleasing to the hunters. Up to 1959 Maurice Burton had sought to emphasize the weight of evidence for undiscovered animals in Loch Ness; in 1960 he began to write rather about the deficiencies of that evidence and canvassed the possible occurrence of gas-propelled mats of vegetation and possibly deceptive sightings of otters and the like. Allegations were made that the O'Connor photograph of 1960 was not genuine. Beppo the Clown dived into Loch Ness as a publicity stunt and generated the canard that a tentacle had grasped at his leg in the water. The Air Force report on Dinsdale's film was misread by a number of people, who then wrote that the film showed an object ninety-two feet long, a crass error that continues to crop up in the literature. A whale bone misappropriated from a museum was deposited for later discovery on the shore of Loch Ness. In 1970 the newspapers made much of a suggestion that Nessie had been killed by pollution of the loch. On April Fool's Day of 1972 a dead elephant seal was floated onto the loch. In 1973 there was a much publicized Japanese expedition that accomplished nothing. Frank Searle had set up as monster-hunter-in-residence in 1969; he published many photos of purported Nessies from 1972 on and produced a newsletter beginning in 1974. Searle's activities have been widely deplored by most of the serious explorers at the loch.

1977 TO DATE

The hunt continues; but there is less publicity and what there is of it tends to be less favorable. The Academy of Applied Science came close to carrying out a search with sonar-triggered cameras and strobes mounted on dolphins, but one of the trained dolphins died before it could be brought to Loch Ness. The Loch Ness and Morar Project obtained support from the Loch Ness

Monster Exhibition at Drumnadrochit and recorded many strong sonar echoes in the deeper waters of the loch. The Goodyear blimp carried scientists and hunters for some fifteen hours of aerial observation over Loch Ness in June 1982. In 1983 an array of sonar transducers was deployed in Urquhart Bay by another U.S. team: had a Nessie-like shape appeared, biopsy darts would have been fired at it. Visitors can now charter a boat (the New Atlantis), equipped with sonar, to try their luck at Nessie hunting.

In 1977 Doc Shiels, stage performer and psychic, set out to bring Nessie to the surface through psychic means; he obtained photos which he has attested by signed deposition to be genuine. In 1983 Erik Beckjord claimed to have filmed several Nessies and wakes, but the results have yet to be published. At the end of 1983 Frank Searle sent out his last newsletter: he was leaving Loch Ness to hunt for buried treasure instead.

During these years several people variously claimed that Nessie is actually an elephant, or the result of atmospheric refraction and distortion, or logs of Scotch pine whose resin is just right for the purpose. In 1983 Binns published the second (after Burton's of 1961) book-length attempt to dismiss the evidence for Nessies, and Steuart Campbell has followed with another. There was also a controversy within a controversy, about the authenticity of the "flipper" shapes photographed in 1972. In 1984 Tim Dinsdale was in the twenty-fifth year of his quest and had mounted his fiftieth personal expedition to the Loch.

By the end of 1985, the only unusual creature positively identified in Loch Ness was a seal which entered the loch toward the end of 1984 and was reportedly shot in late summer of 1985. But the hunt for Nessie had paid off in other directions. A Wellington bomber that had crashed during World War II had been located by sonar during the searches for the monster, and it was salvaged in September 1985 from a depth of 230 feet.[10]

Notes

1. Karl Blind, "Scottish, Shetlandic, and German Water Tales," *Littell's Living Age,* 35 (1881): 811; 36 (1881): 34–36; Forbes Leslie, *The Early Races of Scotland and Their Monuments,* Edinburgh: Edmonston and

Douglas, 1886, pp. 145–46; Alexander Stewart, *'Twixt Ben Nevis and Glencoe,* Edinburgh: William Paterson, 1885, pp. 39–41.

2. Osgood Hanbury MacKenzie, *A Hundred Years in the Highlands,* London: Edward Arnold, 1924, pp. 235–37; *The Journal of Sir Walter Scott, from the Original Manuscript at Abbotsford,* vol. 2, Edinburgh: David Douglas, 1890, pp. 71–72.

3. James M. Mackinlay, *Folklore of Scottish Lochs and Springs,* Glasgow: William Hodge, 1893, p. 173; J. M. McPherson, *Primitive Beliefs in the North-East of Scotland,* London: Longmans, Green, 1929, pp. 62, 69–70.

4. William Mackay, *Urquhart and Glenmoriston—Olden Times in a Highland Parish,* 2d ed., Inverness: Northern Counties Newspaper and Printing and Publishing, 1914, p. 170; W. T. Kilgour, *Lochaber in War and Peace,* Paisley (Eng.): A. Gardner, 1908, pp. 173–74, quoted in Mackinlay, *Folklore of Scottish Lochs,* pp. 247–48.

5. *Inverness Courier,* letter from "Piscator," 29 Aug. 1930; 1933, issues for 2 May, 3 Oct. (p. 5iv–v), 10 Oct. (p. ?v–vi), 20 Oct. (pp. 5vi, 6i), 22 Dec. (p. 5ii–iii); 1934, issues for 9 Jan. (pp. 4ii–iii, 4v–vi, 3i, 5iii–iv), 12 Jan. (pp. 3i, 5iii–iv), 13 Feb. (pp. 3vi, 5ii); *Northern Chronicle,* 27 Aug. 1930; letters from "Camper," "Not an Angler," "Invernessian," 3 Sept. 1930; 11 Oct. 1933; 31 Jan. 1934; *Northern Daily Mail,* 10 Aug. 1934; *Scotsman,* 17, 20 Oct. 1933; *Times* (London), 13 Dec. 1933, p. 10; *Weekly Scotsman,* 10 Feb. 1934. On the lack of a Loch Ness tradition, see *Inverness Courier,* 12 May 1933 (p. 5v), 3 Oct. 1933 (p. 5iv–v), 16 Jan. 1934 (pp. 4v–vi, 5ii), 20 Feb. 1934 (p. 5vi).

6. Rupert T. Gould, "The Loch Ness 'Monster'—A Survey of the Evidence—Fifty-One Witnesses," *Times* (London), 9 Dec. 1933, pp. 13–14; Rupert T. Gould, *The Loch Ness Monster and Others,* New York: University Books, 1969.

7. Personal communication from Hannah Longrigg for Hamish Hamilton, Ltd., 1981.

8. Both the Cockrell and O'Connor photos appeared in Tim Dinsdale, *Loch Ness Monster,* London: Routledge and Kegan Paul, 1961; Roy P. Mackal, *The Monsters of Loch Ness,* Chicago: Swallow, 1976.

9. On the change in tone, see Henry H. Bauer, "Society and Scientific Anomalies: Common Knowledge about the Loch Ness Monster," *Transactions of the Society for Scientific Exploration,* forthcoming.

10. *Aberdeen Press and Journal,* 23 Sept. 1985.

APPENDIX B

Reported Sightings at Loch Ness

A list of reported sightings of the Loch Ness monster can serve a number of purposes: documenting how frequent such reports have been, for example, and whether the frequency has changed or remained the same over long periods; and indicating how many of the reports can be satisfactorily explained without postulating the existence of Nessies. Until recently, the only list available was one published by Roy Mackal in *The Monsters of Loch Ness* of about 250 sightings that he judged valid; now Ulrich Magin has culled the literature for all reported sightings and has generously given permission to publish here a checklist based on his work. Magin's list includes both location of the sighting in the loch and explication of a sighting where known (e.g., hoax); to conserve space I have compressed the list to show only date, name of observer (where known), and published source of the report. An asterisk indicates that a photograph or film was taken.

565

St. Columba: obs. no. 1 in Roy P. Mackal, *The Monsters of Loch Ness,* Chicago: Swallow, 1976; Ronald Binns, *The Loch Ness Mystery Solved,* Shepton Mallet (Somerset): Open Books, 1983, p. 53 and passim; Tim Dinsdale, *Loch Ness Monster,* 4th ed., London: Routledge and Kegan Paul, 1982, p. 27 (all subsequent references are to this edition).

1520

Nicholas Witchell, *The Loch Ness Story,* 2d ed., Lavenham (Suffolk): Terence Dalton, 1976, p. 26; Binns, *Loch Ness Mystery Solved,* 51

1694

R. Franck: Witchell, *Loch Ness Story,* 1976, p. 26; Binns, *Loch Ness Mystery Solved,* 57

17th cent.

Witchell, *Loch Ness Story,* 1976, p. 26

1726

Ibid.; Binns, *Loch Ness Mystery Solved,* 51

1755

Nov. 1: I. Finlay, *The Central Highlands,* Surrey: Batsford, 1976, p. 196

1771

Witchell, *Loch Ness Story,* 1976, p. 26; Binns, *Loch Ness Mystery Solved,* 51; Peter Costello, *In Search of Lake Monsters,* London: Garnstone, 1974, p. 27

1818

MacGruer: Witchell, *Loch Ness Story,* 1976, p. 31

1840s

Grieve: ibid., 36

1852

Inverness Courier, 1 July 1852

1860

Costello, *In Search of Lake Monsters,* 27

1862/65

Jimmy Hossack: ibid., 27, 123

1871/72

D. MacKenzie: obs. no. 2 in Mackal, *Monsters of Loch Ness*

1878

Costello, *In Search of Lake Monsters,* 27; *Inverness Courier,* 12, 20 Feb. 1934

1879

Land sight. no. 2 in Mackal, *Monsters of Loch Ness;* Tim Dinsdale, *The Leviathans,* London: Routledge and Kegan Paul, 1966, p. 56

1880

D. McDonald: Costello, *In Search of Lake Monsters,* 123; Witchell, *Loch Ness Story,* 1976, p. 29

E. Bright: land sight. no. 3 in Mackal, *Monsters of Loch Ness,* 123; Dinsdale, *Leviathans,* 55

1885

R. Matheson: obs. no. 3 in Mackal, *Monsters of Loch Ness;* Costello, *In Search of Lake Monsters,* 28

1889

A. McDonald: obs. no. 4 in Mackal, *Monsters of Loch Ness;* Costello, *In Search of Lake Monsters,* 28; Witchell, *Loch Ness Story,* 1976, pp. 28, 46

Craig: Costello, *In Search of Lake Monsters,* 28; Witchell, *Loch Ness Story,* 1976, p. 205

1880s

Inverness Courier, 20 Oct. 1933

1895

Obs. no. 5 in Mackal, *Monsters of Loch Ness;* Costello, *In Search of Lake Monsters,* 28; Witchell, *Loch Ness Story,* 1976, p. 29

1896

Nov., report in the *Atlanta Constitution* according to John A. Keel, *Strange Creatures from Time and Space,* Greenwich, Conn.: Fawcett, 1970

Report in the *Glasgow Evening News* according to Costello, *In Search of Lake Monsters,* p. 28

James Rose: Witchell, *Loch Ness Story,* 1976, p. 29

1890s

Land sight. no. 4 in Mackal, *Monsters of Loch Ness;* Costello, *In Search of Lake Monsters,* 121

1900

Witchell, *Loch Ness Story,* 1976, p. 131

1903

Dec., F. Fraser: obs. no. 6 in Mackal, *Monsters of Loch Ness;* Costello, *In Search of Lake Monsters,* 29

1908

J. McLeod: *Times* (London), 13 Dec. 1933; obs. no. 7 in Mackal, *Monsters of Loch Ness;* Costello, *In Search of Lake Monsters,* 29

1909

Sept., Mrs. Cameron: land sight. nos. 5–6 in Mackal, *Monsters of Loch Ness;* Costello, *In Search of Lake Monsters,* 33; Witchell, *Loch Ness Story,* 1976, p. 129

1914

July, Mrs. William Miller: obs. no. 8 in Mackal, *Monsters of Loch Ness;* Costello, *In Search of Lake Monsters,* 30

July ?: *Inverness Courier,* 9 Jan. 1934

1915/16

Northern Chronicle, 1 Mar. 1934

1916

James Cameron: obs. 9 in Mackal, *Monsters of Loch Ness;* Costello, *In Search of Lake Monsters,* 29

Andrew Urquhart: *Nessletter,* no. 28, Ness Information Service: R. R. Hepple, Huntshieldford, St. Johns Chapel, Bishop Auckland, Co. Durham, England DL13 1RQ

1917

W. Cary: Dinsdale, *Leviathans,* 195; *National Geographic,* June 1977

1919

Feb., Jack Forbes: Witchell, *Loch Ness Story,* 1976, p. 130

1923

Apr., A. Cruikshank: land sight. no. 7 in Mackal, *Monsters of Loch Ness;* Witchell, *Loch Ness Story,* 1976, p. 131

May 10, W. Miller: obs. no. 10 in Mackal, *Monsters of Loch Ness;* Costello, *In Search of Lake Monsters,* 30

1924

Angus Cameron: *Inverness Courier,* 4 Sept. 1964

1926

Simon Cameron: Witchell, *Loch Ness Story,* 1976, p. 30

1927

Costello, *In Search of Lake Monsters,* 25

F. Macleod: *Northern Chronicle,* 20 Feb. 1935

July, Francis Stuart: *Scotsman,* 16 June 1956

1929

Aug., Mrs. Cummings: obs. no. 11 in Mackal, *Monsters of Loch Ness;* Costello, *In Search of Lake Monsters,* 24

1920s

Witchell, *Loch Ness Story,* 1976, p. 133

1930

July 14 or 22, Ian Milne: obs. no. 12 in Mackal, *Monsters of Loch Ness;* Costello, *In Search of Lake Monsters,* 24; Witchell, *Loch Ness Story,* 1976, p. 46

1931

J. J. Christie: *Inverness Courier,* 5 Mar. 1948

1932

Feb. 7, John Cameron: obs. no. 15 in Mackal, *Monsters of Loch Ness;* Costello, *In Search of Lake Monsters,* 30; Constance Whyte, *More than a Legend,* London: Hamish Hamilton, 1957, p. 204 (all subsequent references are to this edition).

Mrs. MacDonald: obs. no. 16 in Mackal, *Monsters of Loch Ness; Inverness Courier,* 12 Jan. 1934

1933

Inverness Courier, 22 Dec. 1933

Jan., Alexander Davidson: ibid., 12 Jan. 1934

Mar., Miss McDonald: *Times* (London), 15 Dec. 1933

Apr., Sir Compton MacKenzie: Costello, *In Search of Lake Monsters,* 22

Apr. 10: *Reader's Digest* (German edition), Apr. 1977

Apr. 14, Mrs. MacKay: obs. nos. 18–19 in Mackal, *Monsters of Loch Ness;* Witchell, *Loch Ness Story,* 1976, p. 43; Whyte, *More than a Legend,* 206

Spring, A. Davidson: *Inverness Courier,* 12 Jan. 1934

May 11, A. Shaw: obs. no. 21 in Mackal, *Monsters of Loch Ness;* Costello, *In Search of Lake Monsters,* 30

May ?, A.S.: obs. no. 22 in Mackal, *Monsters of Loch Ness;* Whyte, *More than a Legend,* 206

May 27, Mr. & Mrs. Simpson: obs. nos. 23–25 in Mackal, *Monsters of Loch Ness;* Costello, *In Search of Lake Monsters,* 31; Whyte, *More than a Legend,* 205

May ?: Costello, *In Search of Lake Monsters,* 31

May 31: Binns, *Loch Ness Mystery Solved,* 19; *Inverness Courier,* 2 June 1933

May/June, McLennans: obs. no. 26 in Mackal, *Monsters of Loch Ness;* Costello, *In Search of Lake Monsters,* 30

June 1: Binns, *Loch Ness Mystery Solved,* 19; Maurice Burton, *The Elusive Monster,* London: Rupert Hart-Davis, 1961, p. 19; Costello, *In Search of Lake Monsters,* 31

June 8, J. Cameron: Binns, *Loch Ness Mystery Solved,* 88; *Inverness Courier,* 2 June 1933; Witchell, *Loch Ness Story,* 1976, p. 47

June 15, F. Sutherland: *Northern Chronicle,* 21 June 1933

June 27: Witchell, *Loch Ness Story,* 1976, p. 47

June ?, A. Ross: obs. no. 27 in Mackal, *Monsters of Loch Ness*

July 1, Lay monk: *Northern Chronicle,* 12 July 1933

July 8: *Northern Chronicle,* 12 July 1933

July 10, A. Frazer: *Zeit,* 31 Dec. 1936

July 15, J. Simpson: obs. no. 28 in Mackal, *Monsters of Loch Ness*

July 22, Spicer: *Inverness Courier,* 4 Aug. 1933; land sight. no. 11 in Mackal, *Monsters of Loch Ness;* Dinsdale, *Loch Ness Monster,* 32; Burton, *Elusive Monster,* 149

July 26, Wm. Grant: *Northern Chronicle,* 26 July 1933

Aug., Mrs. McLennan: land sight. no. 12 in Mackal, *Monsters of Loch Ness;* Burton, *Elusive Monster,* 150; Costello, *In Search of Lake Monsters,* 35; Witchell, *Loch Ness Story,* 1976, p. 136

Aug. 5, Miss N. Smith: obs. no. 29 in Mackal, *Monsters of Loch Ness;*

Binns, *Loch Ness Mystery Solved,* 21, 173; *Inverness Courier,* 8 Aug. 1933

Aug. 5, Miss Keyes: obs. no. 30 in Mackal, *Monsters of Loch Ness;* Costello, *In Search of Lake Monsters,* 41; *Inverness Courier,* 8 Aug. 1933

Aug. 9, Mrs. Cheshire: Binns, *Loch Ness Mystery Solved,* 22; *Inverness Courier,* 11 Aug. 1933

Aug. 11, A. H. Palmer: obs. no. 31 in Mackal, *Monsters of Loch Ness;* Costello, *In Search of Lake Monsters,* 41; Whyte, *More than a Legend,* 82

Aug. 15, John Cameron: obs. no. 34 in Mackal, *Monsters of Loch Ness*

Aug. 16, Mrs. Scott: obs. no. 36 in ibid.; Whyte, *More than a Legend,* 205

Aug. 18, Mrs. McDonnell: obs. no. 37 in Mackal, *Monsters of Loch Ness;* Whyte, *More than a Legend,* 205

Aug. 19, Moir: obs. no. 38 in Mackal, *Monsters of Loch Ness*

Aug. 24, Mrs. Rathray: Burton, *Elusive Monster,* 24

Aug. 26, W.D.H.M.: Whyte, *More than a Legend,* 204

Aug. 31: *Inverness Courier,* 6 Sept. 1933

Aug. ?, Hector Macphail: obs. no. 41 in Mackal, *Monsters of Loch Ness*

Aug. ?, G. McQueen: obs. no. 32 in ibid.; Whyte, *More than a Legend,* 205

Aug. ?, R. Fullerton: obs. no. 33 in Mackal, *Monsters of Loch Ness*

Aug. ?, A.J.G.: obs. no. 35 in ibid.; Whyte, *More than a Legend,* 202

Sept. 13, J. Cameron: *Inverness Courier,* 15 Sept. 1933

Sept. 14: ibid.

Sept. 14, James Cameron: ibid.; Binns, *Loch Ness Mystery Solved,* 22

Sept. 22, Miss Fraser: obs. no. 43 in Mackal, *Monsters of Loch Ness;* Burton, *Elusive Monster,* 119; Costello, *In Search of Lake Monsters,* 42; Witchell, *Loch Ness Story,* 1976, p. 48

Sept. 22, D. W. Morrison: obs. nos. 44, 46 in Mackal, *Monsters of Loch Ness;* Costello, *In Search of Lake Monsters,* 42; Whyte, *More than a Legend,* 204

Sept. 22, George Macqueen: *Inverness Courier,* 26 Sept. 1933

Sept. 22, A. Campbell: Whyte, *More than a Legend,* 203; obs. no. 45 in Mackal, *Monsters of Loch Ness;* Binns, *Loch Ness Mystery Solved,* 76

Sept. ?, James Cameron: Whyte, *More than a Legend,* 204; obs. no. 47 in Mackal, *Monsters of Loch Ness*

Sept. ?, J. M. McSkimming: obs. no. 42 in Mackal, *Monsters of Loch Ness*

Oct. 1, Russell: obs. no. 48 in ibid.; Binns, *Loch Ness Mystery Solved,* 122; Costello, *In Search of Lake Monsters,* 43

Oct. 18, Mrs. G.: obs. no. 49 in Mackal, *Monsters of Loch Ness;* Whyte, *More than a Legend,* 202

Oct. 20, Connell (Scott II): obs. no. 50 in Mackal, *Monsters of Loch Ness; Inverness Courier,* 24 Oct. 1933

Oct. 22, Mrs. J. Simpson: obs. no. 51 in Mackal, *Monsters of Loch Ness;* Costello, *In Search of Lake Monsters,* 43; Binns, *Loch Ness Mystery Solved,* 122
Oct. 22, A. Gillies: *Inverness Courier,* 27 Oct. 1933; Whyte, *More than a Legend,* 201–2; obs. no. 52 in Mackal, *Monsters of Loch Ness*
Oct. 22 or 27, Miss McDonald: obs. no. 54 in Mackal, *Monsters of Loch Ness*
Oct. 30, McLeod: obs. no. 55 in ibid.; *Inverness Courier,* 31 Oct. 1933
Oct. ?, Whyte, *More than a Legend,* 49; Costello, *In Search of Lake Monsters,* 66
Autumn, O'Flynn: obs. no. 56 in Mackal, *Monsters of Loch Ness*
Nov. 10, McRae: obs. no. 57 in ibid.; *Inverness Courier,* 14 Dec. 1933
Nov. 10, Mrs. A. Pimley: *Inverness Courier,* 14 Dec. 1933
Nov. 11, John Cameron: obs. no. 58 in Mackal, *Monsters of Loch Ness;* Whyte, *More than a Legend,* 201
*Nov. 12, Hugh Gray: photo 1 in Mackal, *Monsters of Loch Ness;* Burton, *Elusive Monster,* 78; Costello, *In Search of Lake Monsters,* 38; Witchell, *Loch Ness Story,* 1976, p. 56
Nov. 14, Mr. & Mrs. Kirton: obs. no. 59 in Mackal, *Monsters of Loch Ness;* Binns, *Loch Ness Mystery Solved,* 93; *Times* (London), 27 Dec. 1933
mid-Nov., B. MacKenzie: *Times* (London), 18 Dec. 1933
Nov. 17, A.R.McF.: obs. no. 60 in Mackal, *Monsters of Loch Ness;* Whyte, *More than a Legend,* 201
Nov. 20, Miss Simpson: Burton, *Elusive Monster,* 122; obs. no. 61 in Mackal, *Monsters of Loch Ness*
Nov. 23: *Times* (London), 9 Dec. 1933
Nov. 24, Rev. D.: obs. no. 62 in Mackal, *Monsters of Loch Ness*
Oct./Nov., Mrs. Cameron: *Inverness Courier,* 12 Jan. 1934
early Dec., Mr. & Mrs. MacKenzie: *Times* (London), 18 Dec. 1933
Dec. 6, A. Macdonald: ibid.
Dec. 6, A. Bisset: *Inverness Courier,* 8 Dec. 1933
Dec. 11, Mr. O.: obs. no. 63 in Mackal, *Monsters of Loch Ness;* Whyte, *More than a Legend,* 201
*Dec. 12, M. Irvine: film 3 in Mackal, *Monsters of Loch Ness;* Burton, *Elusive Monster,* 72; Costello, *In Search of Lake Monsters,* 43; Witchell, *Loch Ness Story,* 1976, p. 57
Dec. 13, A. Gillies: *Inverness Courier,* 15 Dec. 1933
Dec. 18: *Times* (London), 18 Dec. 1933
Dec. 19: *Daily Mail,* 21 Dec. 1933
Dec. 20, Wetherall: Costello, *In Search of Lake Monsters,* 45; Witchell, *Loch Ness Story,* 1976, p. 60
Dec. 20, Mrs. Reid: land sight. no. 12 in Mackal, *Monsters of Loch Ness;* Costello, *In Search of Lake Monsters,* 35; Witchell, *Loch Ness Story,* 1976, p. 136; *Inverness Courier,* 26 Dec. 1933

Dec. 25, Mach: obs. no. 64 in Mackal, *Monsters of Loch Ness;* Whyte, *More than a Legend,* 205
Dec. 26, J. Kirton: obs. no. 65 in Mackal, *Monsters of Loch Ness; Times* (London), 27 Dec. 1933
Dec. 27, G. Jamieson: obs. no. 66 in Mackal, *Monsters of Loch Ness;* Burton, *Elusive Monster,* 123; Whyte, *More than a Legend,* 200
Dec. 27, S. Mackintosh: *Times* (London), 29 Dec. 1933
Dec. 28: ibid.
Dec. 29: *Inverness Courier,* 29 Dec. 1933
Dec. 30, John Deans: *Times* (London), 1 Jan. 1934; obs. no. 67 in Mackal, *Monsters of Loch Ness*
Dec. 30, Goodbody: obs. no. 68 in Mackal, *Monsters of Loch Ness*
Dec. ?, H. F. Hay: Costello, *In Search of Lake Monsters,* 48
?, Mrs. E. Price-Hughes: land sight. no. 10 in Mackal, *Monsters of Loch Ness;* Costello, *In Search of Lake Monsters,* 121
?, Wm. McLean: Witchell, *Loch Ness Story,* 1976, p. 136
?, Mrs. Kirton: obs. no. 17 in Mackal, *Monsters of Loch Ness*
?, A. J. Gray: *Inverness Courier,* 21 June 1935
?: Witchell, *Loch Ness Story,* 1976, p. 53

1934

Costello, *In Search of Lake Monsters,* 143
early, Mrs. McL.: obs. no. 70 in Mackal, *Monsters of Loch Ness;* Whyte, *More than a Legend,* 200
early Jan., Mrs. Goodbody: obs. no. 69 in Mackal, *Monsters of Loch Ness*
Jan. 2: *Inverness Courier,* 5 Jan. 1934; *Northern Chronicle,* 17 Jan. 1934
Jan. 3: *Inverness Courier,* 5 Jan. 1934
Jan. 3, Wm. Macintosh: *Times* (London), 4 Jan. 1934
Jan. 4: "Our Special Artist Investigates the Loch Ness Monster," *Illustrated London News,* 13 Jan. 1934, pp. 1, 40–41
Jan. 4, A. Grant: land sight. no. 14 in Mackal, *Monsters of Loch Ness;* Costello, *In Search of Lake Monsters,* 46; Witchell, *Loch Ness Story,* 1976, p. 137; Burton, *Elusive Monster,* 146
Jan. 4, Alexander Rose: *Daily Mail,* 5 Jan. 1934
Jan. 7, Lind Davidson: *Inverness Courier,* 9 Jan. 1934
Jan. 7: *Daily Mail,* 8 Jan. 1934
Jan. 8, A. Davidson: *Inverness Courier,* 12 Jan. 1934
Jan. 10, J.H.F., M.A.H.: obs. no. 71 in Mackal, *Monsters of Loch Ness;* Whyte, *More than a Legend,* 199
Jan. 10, G. Jamieson: unidentified sources
Jan. 11, Mrs. Tinnock: *Inverness Courier,* 12 Jan. 1934
Jan. 13: ibid., 16 Jan. 1934
Jan. 15: ibid.
Jan. 15, Wetherall: Witchell, *Loch Ness Story,* 1976, p. 63

Jan. 30, Carson: obs. no. 72 in Mackal, *Monsters of Loch Ness;* Burton, *Elusive Monster,* 123; Whyte, *More than a Legend,* 198; *Inverness Courier,* 2 Feb. 1934
Feb. 9, D. Williamson: *Inverness Courier,* 13 Feb. 1934
Feb. 11, D. Grant: ibid.
Feb. 22, Mrs. McDonald: obs. no. 73 in Mackal, *Monsters of Loch Ness; Inverness Courier,* 27 Feb. 1934
Feb. 24, Mrs. Hill: *Inverness Courier,* 27 Feb. 1934; Whyte, *More than a Legend,* 198
Feb. 25, Mrs. McLennan: ibid.; obs. no. 74 in Mackal, *Monsters of Loch Ness;* Burton, *Elusive Monster,* 123
Feb. ?, P. Harvey: land sight. no 15 in Mackal, *Monsters of Loch Ness;* Costello, *In Search of Lake Monsters,* 49; Witchell, *Loch Ness Story,* 1976, p. 139
Mar. 2, J. Simpson: *Northern Chronicle,* 7 Mar. 1934
Mar. 3, R. Mitchell: *Inverness Courier,* 6 Mar. 1934
Mar. 3, J. Campbell: ibid.
Mar. 5, J. Macdonald: ibid.
Mar. 28, I. Finkelstein: ibid., 30 Mar. 1934
Apr. 1: ibid., 3 Apr. 1934
Apr. 1, Miss Warwick: ibid.; obs. no. 75 in Mackal, *Monsters of Loch Ness*
Apr. 1, D. Fraser: *Inverness Courier,* 3 Apr. 1934
Apr. 1, G. Rodgers: ibid.
Apr. 2: *Northern Chronicle,* 4 Apr. 1934
Apr. 16, J. Hunt: *Inverness Courier,* 17 Apr. 1934; obs. no. 76 in Mackal, *Monsters of Loch Ness*
Apr. 16, Macdonald: *Inverness Courier,* 17 Apr. 1934
*Apr. 19, K. Wilson: ibid., 20 Apr. 1934; photo 2 in Mackal, *Monsters of Loch Ness;* Witchell, *Loch Ness Story,* 1976, p. 65
Apr.?, Major J.H.: Whyte, *More than a Legend,* 198
May 1, Miss K. McDonald: obs. no. 77 in Mackal, *Monsters of Loch Ness;* Burton, *Elusive Monster,* 123; Whyte, *More than a Legend,* 197
May 4, Kenneth: *Inverness Courier,* 8 May 1934
May 7, K. Cameron: *Northern Chronicle,* 9 May 1934
May 7, G. Grinton: *Inverness Courier,* 8 May 1934
May 18, Dyer: ibid., 18 May 1934
May 20, H. Matheson: *Northern Chronicle,* 23 May 1934; *Daily Mail,* 23 May 1934
May 24, H. Ross: *Inverness Courier,* 25 May 1934
May 26, Brother Horan: obs. no. 79 in Mackal, *Monsters of Loch Ness;* Burton, *Elusive Monster,* 129; Costello, *In Search of Lake Monsters,* 66; Witchell, *Loch Ness Story,* 1976, p. 85
May 30, Miss Fraser: *Inverness Courier,* 1 June 1934
May 31, J. Macdonald: ibid., 5 June 1934

June 3, C. MacKintosh: ibid.
June 3: *Northern Chronicle,* 6 June 1934
June 5, Mrs. Munro: land sight. no. 16 in Mackal, *Monsters of Loch Ness;* Burton, *Elusive Monster,* 145; Witchell, *Loch Ness Story,* 1976, p. 140; Costello, *In Search of Lake Monsters,* 67; Binns, *Loch Ness Mystery Solved,* 91
June 9, J. Peter: *Northern Chronicle,* 13 June 1934
*June 10: photo 3 in Mackal, *Monsters of Loch Ness;* Burton, *Elusive Monster,* 81; Costello, *In Search of Lake Monsters,* 65; Witchell, *Loch Ness Story,* 1976, p. 77
June 11, J. Dean: *Inverness Courier,* 12 June 1934
June 16, D. Petrie: Witchell, *Loch Ness Story,* 1976, p. 83; *Times* (London), 18 June 1934
June 24, Miss Fraser: Costello, *In Search of Lake Monsters,* 68; obs. no. 80 in Mackal, *Monsters of Loch Ness*
June 25, Ch. Macdonald: *Inverness Courier,* 26 June 1934
June 25, M. Macmillan: ibid.
June 26, J. Fraser: *Northern Chronicle,* 27 June 1934
June 29, J. Harting: *Inverness Courier,* 3 July 1934
June 30: ibid.; Costello, *In Search of Lake Monsters,* 68; obs. no. 81 in Mackal, *Monsters of Loch Ness;* Whyte, *More than a Legend,* 47
June ?, John Young: Witchell, *Loch Ness Story,* 1976, p. 75
July 4: Costello, *In Search of Lake Monsters,* 61; A. Hall, *Bestien, Scheusale und Monster,* Berlin: Ullstein, 1979, p. 76
July 12, F. Haselfoot: Burton, *Elusive Monster,* 127; Costello, *In Search of Lake Monsters,* 56; *Times* (London), 16 July 1934
July 12, Chambers: *Inverness Courier,* 13 July 1934
July 12, William McKay: Burton, *Elusive Monster,* 51; obs. no. 82 in Mackal, *Monsters of Loch Ness*
July 12, R. J. Scott: Burton, *Elusive Monster,* 49; *Inverness Courier,* 13 July 1934; obs. no. 85 in Mackal, *Monsters of Loch Ness*
*July 13: photo 4 in Mackal, *Monsters of Loch Ness*
July 14, R. J. Scott: obs. no. 86 in ibid.; Burton, *Elusive Monster,* 52
July 14, A. Macmillan: *Inverness Courier,* 17 July 1934
July 15, Miss Pillon: ibid.
July 16, Macintosh: ibid.; Burton, *Elusive Monster,* 52; obs. no. 87 in Mackal, *Monster of Loch Ness*
July 16, Mrs. Briddle: *Inverness Courier,* 17 July 1934
July 17, Ross: Burton, *Elusive Monster,* 53; obs. no. 88 in Mackal, *Monsters of Loch Ness*
July 18, W. McKay: Burton, *Elusive Monster,* 53; obs. no. 89 in Mackal, *Monsters of Loch Ness*
July 18, W. Campbell: Burton, *Elusive Monster,* 57
July 19, McKay: ibid., 54; obs. no. 90 in Mackal, *Monsters of Loch Ness*
July 19, G. Lobban: *Northern Chronicle,* 25 July 1934
July 19, C. Graham: ibid.

July 22, William MacDonald: ibid.
July 24, Ch. Mace: ibid.
July 24: ibid.
July 24, R. Urquhart: Burton, *Elusive Monster,* 55
July 24, Ralph: ibid., 56; obs. no. 91 in Mackal, *Monsters of Loch Ness*
July 24, Duncan Cameron: *Inverness Courier,* 27 July 1934
July 25, W. Campbell: ibid., 31 July 1934; Burton, *Elusive Monster,* 56; obs. no. 92 in Mackal, *Monsters of Loch Ness*
July 26, John Tolmie: *Inverness Courier,* 31 July 1934
July 27, Grant: Burton, *Elusive Monster,* 56, 140; obs. no. 93 in Mackal, *Monsters of Loch Ness*
July 30, Scott: Burton, *Elusive Monster,* 57; obs. no. 95 in Mackal, *Monsters of Loch Ness*
July 30, W. Campbell: Burton, *Elusive Monster,* 57, 140; Costello, *In Search of Lake Monsters,* 57; obs. no. 94 in Mackal, *Monsters of Loch Ness*
July ?, E.F.: obs. no. 96 in Mackal, *Monsters of Loch Ness;* Whyte, *More than a Legend,* 197
July ?, Mrs. Cameron: Costello, *In Search of Lake Monsters,* 34
July ?, Ian J. Matheson: Whyte, *More than a Legend,* 196; Witchell, *Loch Ness Story,* 1976, p. 141
July ?: Burton, *Elusive Monster,* 55
July ?, W.McL.C.: Whyte, *More than a Legend,* 197
July ?, W. Milligan: *Northern Chronicle,* 1 Aug. 1934
July 30–Aug. 4, Lawson: *Inverness Courier,* 7 Aug. 1934
Aug. 1, Mrs. Smith: ibid., 3 Aug. 1934
Aug. 1, J. Hartwell: ibid.
*Aug. 5, F. C. Adams: Burton, *Elusive Monster,* 80; Costello, *In Search of Lake Monsters,* 64; photo 5 in Mackal, *Monsters of Loch Ness*
Aug. 5, J. Fraser: *Inverness Courier,* 7 Aug. 1934
Aug. 5, J. Rigg: ibid.
Aug. 7, Capt. Fraser: Burton, *Elusive Monster,* 59; obs. no. 99 in Mackal, *Monsters of Loch Ness*
Aug. 7, Mr. Alexander: *Northern Chronicle,* 8 Aug. 1934
Aug. 8, Sir Macdonald: *Inverness Courier,* 14 Aug. 1934; obs. no. 100 in Mackal, *Monsters of Loch Ness;* Whyte, *More than a Legend,* 47; Witchell, *Loch Ness Story,* 1976, p. 83
Aug. 11, I. Kirton: *Inverness Courier,* 14 Aug. 1934
Aug. 11, P.G.: Whyte, *More than a Legend,* 196
Aug. 12, Grant: Burton, *Elusive Monster,* 60; Costello, *In Search of Lake Monsters,* 58; obs. no. 101 in Mackal, *Monsters of Loch Ness;* Witchell, *Loch Ness Story,* 1976, p. 70
Aug. 12, H. Schields: *Inverness Courier,* 14 Aug. 1934
Aug. 13, D. J. Peterson: ibid., 17 Aug. 1934
mid-Aug., I. Armitage: *Northern Chronicle,* 15 Aug. 1934
Aug. 20, George Somerset: *Inverness Courier,* 24 Aug. 1934

Aug. 26, Mrs. MacKinnon: ibid., 28 Aug. 1934
Aug. 30, Dom Cyril: Costello, *In Search of Lake Monsters,* 66; obs. no. 102 in Mackal, *Monsters of Loch Ness;* Whyte, *More than a Legend,* 87
Summer, J.C.: obs. no. 98 in Mackal, *Monsters of Loch Ness;* Whyte, *More than a Legend,* 197
Sept. 4, The Sisters: *Northern Chronicle,* 5 Sept. 1934
Sept. 7, Miss Fraser: *Inverness Courier,* 11 Sept. 1934
Sept. 7: ibid.
Sept. 9, A. Fraser: ibid.
*Sept. 15, J. Fraser: photo 4 in Mackal, *Monsters of Loch Ness;* Binns, *Loch Ness Mystery Solved,* 39; Witchell, *Loch Ness Story,* 1976, p. 71
Sept. 15: *Northern Chronicle,* 20 Feb. 1935
Sept. 29, Miss Chisholm: *Inverness Courier,* 2 Oct. 1934
Sept. 29, J. Campbell: ibid.
?9: *Northern Chronicle,* 23 Sept. 1936
Oct. 2, E. Cameron: *Northern Chronicle,* 3 Oct. 1934
Oct. 4, Mrs. MacCallum: *Inverness Courier,* 9 Oct. 1934
Oct. 13, Mrs. Grant: ibid., 16 Oct. 1934
Nov. 15, H. Paterson: *Northern Chronicle,* 21 Nov. 1934
Nov. 16, H. Matheson: ibid.
Dec. 6, J. Sinclair: ibid., 12 Dec. 1934
Dec. 27, W. Wotherspoon: ibid., 2 Jan. 1935
late, Mrs. C.A.McG., Miss R.C.: obs. no. 107 in Mackal, *Monsters of Loch Ness;* Whyte, *More than a Legend,* 196
late ?, I.M.: obs. no. 97 in Mackal, *Monsters of Loch Ness*
late ?, H.F., Col. W.: obs. no. 104 in ibid.
late ?, D.McK., J.C.: obs. no. 105 in ibid.
late ?, J.C.: obs. no. 106 in ibid.

1935

Jan. 13, D. Fraser: *Northern Chronicle,* 16 Jan. 1935
Jan. 19: *Inverness Courier,* 22 Jan. 1935; T. Shields, *They Saw Nessie,* Gartochen: Northern Books, 1984, p. 19
Feb. 7, D. Ross: *Inverness Courier,* 8 Feb. 1935
Apr. 19, Mrs. Scott: ibid., 23 Apr. 1935
May 23, George Clark: ibid., 24 May 1935
June 10, Mrs. Magrath: *Northern Chronicle,* 12 June 1935
June 17, R. Forbes: ibid., 19 June 1935
June 19, A. J. Gray: *Inverness Courier,* 21 June 1935
June 21, G. Sutherland: ibid., 25 June 1935
June 23: Whyte, *More than a Legend,* 20
early July, Arthur Pimley: *Northern Chronicle,* 10 July 1935
mid-July, Mr. Clelland: ibid., 24 July 1935
Aug. 2, Rev. Moore: *Inverness Courier,* 6 Aug. 1935

Aug. 4, Maj. Phillips: ibid.
Aug. 16, P. Grant: ibid., 20 Aug. 1935
Aug. 17: ibid.
Aug. 18, Ian MacPhail: ibid.
Aug. 18, M. Steward: *Northern Chronicle,* 21 Aug. 1935
Aug. 19, C. W. Farnham: ibid.
Aug. 20, Count Bentinck: *Inverness Courier,* 20 Aug. 1935
Aug. 24, Count Bentinck: ibid., 10 Sept. 1935; obs. no. 103 in Mackal, *Monsters of Loch Ness;* Costello, *In Search of Lake Monsters,* 65; Witchell, *Loch Ness Story,* 1976, p. 202
Summer, H. Gordon: obs. no. 108 in Mackal, *Monsters of Loch Ness;* Whyte, *More than a Legend,* 61
Sept. 5, R. Stone: *Inverness Courier,* 10 Sept. 1935
Sept. 9, G. Watt: ibid.
Autumn: J. A. Carruth, *Loch Ness and Its Monsters,* 9th ed., Fort Augustus: Abbey Press, 1971, p. 15.
Dec. 24, R. McKenzie: Shields, *They Saw Nessie,* 21; Witchell, *Loch Ness Story,* 1976, p. 85

1936

Mar. 22, Alpin Chisholm: *Inverness Courier,* 24 Mar. 1936; *Northern Chronicle,* 25 Mar. 1936
end Mar., Mrs. Grant: *Inverness Courier,* 31 Mar. 1936
May 10, James Mavor: *Northern Chronicle,* 13 May 1936
June, Mr. & Mrs. Hallam: obs. no. 109 in Mackal, *Monsters of Loch Ness*
June 20: *Northern Chronicle,* 24 June, 1936
June 20: ibid.
June 20, H. A. Frere: ibid.; *Inverness Courier,* 23 June 1936
June 21: *Inverness Courier,* 23 June 1936
June 21, Miss H. MacFie: ibid.
July 23, H. A. Frere: ibid., 28 Sept. 1937
Sept. 9, I. Macintosh: ibid., 15 Sept. 1936; *Northern Chronicle,* 16 Sept. 1936
Sept. 17, C. Campbell: *Inverness Courier,* 22 Sept. 1936; *Northern Chronicle,* 23 Sept. 1936
*Sept. 22, M. Irvine: film 5 in Mackal, *Monsters of Loch Ness;* Costello, *In Search of Lake Monsters,* 71; Witchell, *Loch Ness Story,* 1976, p. 78
Sept. 29: *Northern Chronicle,* 30 Sept. 1936; *Inverness Courier,* 22 Sept. 1936
Sept. 29, Miss Jobson: *Northern Chronicle,* 30 Sept. 1936
Oct. 28, D. McMillan: Witchell, *Loch Ness Story,* 1976, p. 86
Oct. ?, Mrs. Moir: Costello, *In Search of Lake Monsters,* 74; obs. no. 110 in Mackal, *Monsters of Loch Ness*

Nov. ?, H. MacKenzie: *Inverness Courier,* 24 Nov. 1936; *Northern Chronicle,* 25 Nov. 1936

1936/37

June, Jimmie Williams: *Nessletter,* no. 66
Summer, N. LePoer: Burton, *Elusive Monster,* 110

1937

Jan. 3, Farrel: Costello, *In Search of Lake Monsters,* 68; obs. no. 111 in Mackal, *Monsters of Loch Ness*
early Feb., C. A. Macgruer: *Inverness Courier,* 12 Feb. 1937
late Feb., Mr. Campbell: ibid., 2 Mar. 1937
Mar. 15, C. McPherson Young: ibid., 16 Mar. 1937
Easter, A. Erskine Murray: Burton, *Elusive Monster,* 125; Costello, *In Search of Lake Monsters,* 74
Apr. 16: *Inverness Chronicle,* 20 Apr. 1937
Apr. 17: ibid.
Apr. 18, B. Macrae: ibid.
June, A. Smith: Costello, *In Search of Lake Monsters,* 69; obs. no. 112 in Mackal, *Monsters of Loch Ness*
mid-June: *Northern Chronicle,* 30 June 1937
July 13, Gourlay: Costello, *In Search of Lake Monsters,* 69; obs. no. 113 in Mackal, *Monsters of Loch Ness*
July 25, William Young: *Inverness Courier,* 27 July 1937
July 27, L.A.R.: Burton, *Elusive Monster,* 126; Costello, *In Search of Lake Monsters,* 75; obs. no. 114 in Mackal, *Monsters of Loch Ness;* Whyte, *More than a Legend,* 195
July 30, P. Grant: *Inverness Courier,* 3 Aug. 1937
Aug. 12: ibid., 17 Aug. 1937
Aug., Rev. Graham: obs. no. 115 in Mackal, *Monsters of Loch Ness;* Whyte, *More than a Legend,* 49
Summer, Frs. Grimet, McKinnon: Costello, *In Search of Lake Monsters,* 76
Sept. 9, Mrs. Ferguson: *Northern Chronicle,* 11 Sept. 1937
Sept. 18, M.A.F.: *Inverness Courier,* 24 Sept. 1937
Nov. 7, P. Walker: ibid., 9 Nov. 1937
?, D. Hunter: obs. no. 117 in Mackal, *Monsters of Loch Ness*
?, M.: Whyte, *More than a Legend,* 195

1938

Jan., W. MacKay: obs. no. 118 in Mackal, *Monsters of Loch Ness;* Witchell, *Loch Ness Story,* 1976, p. 90
Jan. 26, Wm. M'Lellan: *Inverness Courier,* 28 Jan. 1938
late Feb.: ibid., 1 Mar. 1938
late Feb., D. McDonald: ibid.
Feb. 26: ibid.

Apr. 25, C. B. Prickett: *Times* (London), 9 May 1938; Whyte, *More than a Legend,* 59

*May 29, G. E. Taylor: Burton, *Elusive Monster,* 83; film 6 in Mackal, *Monsters of Loch Ness*

June, McLean: Burton, *Elusive Monster,* 128; Costello, *In Search of Lake Monsters,* 75; obs. nos. 116, 119 in Mackal, *Monsters of Loch Ness*

early June: *Inverness Courier,* 14 June 1938

early June, H. Colvin: ibid.

July 11, Mrs. Steward: ibid., 12 July 1938

Aug. 2, J. J. Forsyth: *Northern Chronicle,* 3 Aug. 1938

Aug. 9, A. MacLaren: ibid., 10 Aug. 1938

Aug. 16, Rev. M'Kinnon: *Inverness Courier,* 19 Aug. 1938; *Northern Chronicle,* 17 Aug. 1938

Aug. 25, Mr. Battle: *Inverness Courier,* 26 Aug. 1938

Aug. 30, W. Brodie: obs. no. 120 in Mackal, *Monsters of Loch Ness;* Whyte, *More than a Legend,* 195

Aug. 31, Wm. Grant: *Inverness Courier,* 2 Sept. 1938

Aug. 31, J. McEwan: ibid.

Summer, Mr. & Mrs. Wotherspoon: obs. no. 121 in Mackal, *Monsters of Loch Ness*

Summer, R. Synge: Witchell, *Loch Ness Story,* 1976, p. 97

?, Anne J. Thomas: *Scots Magazine,* Jan. 1980

*?, J. Carrie: Witchell, *Loch Ness Story,* 1976, p. 206

?, Sir David Hunter Blair: Whyte, *More than a Legend,* 59

?: ibid., 75

?: *Scots Magazine,* May 1962

1939

Jan. 16, H. A. Frere: *Inverness Courier,* 18 Jan. 1939

Apr. 5, L. P. Prover: ibid., 7 Apr. 1939

May 24, J. Macdonald: ibid., 26 May 1939; *Northern Chronicle,* 31 May 1939

June, A. C. Martin: obs. no. 123 in Mackal, *Monsters of Loch Ness;* Whyte, *More than a Legend,* 44; Witchell, *Loch Ness Story,* 1976, p. 96

June 21, Th. Campbell: *Inverness Courier,* 23 June 1939

July 3, W. Macdonald: ibid., 4 July 1939

Aug. 4, Munro: ibid., 8 Aug. 1939

Aug. 5, Gilchrist: ibid.

Aug. 12, W. Fraser: ibid., 15 Aug. 1939

Aug. 16, Duguid: ibid., 18 Aug. 1939

?, S. Hunter Gordon: Costello, *In Search of Lake Monsters,* 69; obs. no. 127 in Mackal, *Monsters of Loch Ness*

?: Dinsdale, *Leviathans,* 56

?, R. McEwen: obs. nos. 124–25 in Mackal, *Monsters of Loch Ness;* Witchell, *Loch Ness Story,* 1976, pp. 87–88
?, W. MacKay: obs. no. 126 in Mackal, *Monsters of Loch Ness*

1930s

D.A.C.: obs. no. 14 in Mackal, *Monsters of Loch Ness*
Alec Muir: land sight. no. 8 in ibid.; Costello, *In Search of Lake Monsters,* 120
Schoolchildren: Burton, *Elusive Monster,* 152; Costello, *In Search of Lake Monsters,* 121; land sight no. 9 in Mackal, *Monsters of Loch Ness;* Witchell, *Loch Ness Story,* 1976, p. 136
early, Alex Campbell: Dinsdale, *Loch Ness Monster,* 114
end: Witchell, *Loch Ness Story,* 1976, p. 82

1940

May 5, Dom Cyril: Costello, *In Search of Lake Monsters,* 66; obs. no. 129 in Mackal, *Monsters of Loch Ness;* Whyte, *More than a Legend,* 87
? J.McD.: obs. no. 128 in Mackal, *Monsters of Loch Ness;* Whyte, *More than a Legend,* 195
?, G. E. Carruth: Witchell, *Loch Ness Story,* 1976, p. 103

1940–45

James Cameron: ibid., 205; obs. no. 130 in Mackal, *Monsters of Loch Ness*

1941

Aug.: Costello, *In Search of Lake Monsters,* 76

1942

June: *Time,* 29 June 1942

1943

C. B. Farrel: obs. no. 131 in Mackal, *Monsters of Loch Ness*
Jan. 8, S. Grant: obs. no. 132 in ibid.; Costello, *In Search of Lake Monsters,* 68
May 25, C. B. Farrel: Burton, *Elusive Monster,* 130; obs. no. 133 in Mackal, *Monsters of Loch Ness;* Witchell, *Loch Ness Story,* 1976, p. 89
?, R. Flint: Witchell, *Loch Ness Story,* 1976, p. 105

1944

Apr., Meiklem: obs. no. 134 in Mackal, *Monsters of Loch Ness;* Whyte, *More than a Legend,* 55; Witchell, *Loch Ness Story,* 1976, p. 104
Summer, J.MacF.B.: obs. no. 135 in Mackal, *Monsters of Loch Ness;* Whyte, *More than a Legend,* 194

1945

May, Mr. & Mrs. Lane: obs. no. 136 in Mackal, *Monsters of Loch Ness;* Whyte, *More than a Legend,* 194

Summer, I.C.: obs. no. 137 in Mackal, *Monsters of Loch Ness;* Whyte, *More than a Legend,* 194
Sept., Gavin Maxwell: Costello, *In Search of Lake Monsters,* 106
?, William MacKay: obs. no. 119 in Mackal, *Monsters of Loch Ness;* Witchell, *Loch Ness Story,* 1976, p. 90

1946

July 28, D. A. Campbell: *Inverness Courier,* 30 July 1946
Aug., N. Atkinson: Burton, *Elusive Monster,* 158; Costello, *In Search of Lake Monsters,* 77

1947

Apr. 4: Costello, *In Search of Lake Monsters,* 77; obs. no. 139 in Mackal, *Monsters of Loch Ness;* Whyte, *More than a Legend,* 40; Witchell, *Loch Ness Story,* 1976, p. 90
Apr. 4, J. W. Killop: *Inverness Courier,* 8 Apr. 1947
June, Brother Mark: *Scotsman,* 16 June 1956
July 18: *Inverness Courier,* 1 Aug. 1947
July 27: ibid.; obs. nos. 140–41 in Mackal, *Monsters of Loch Ness*
July 27, J. C. Forbes: Janet Bord & Colin Bord, *Alien Animals,* Harrisburg, Pa.: Stackpole Books, 1981, p. 110; *Inverness Courier,* 1 Aug. 1947; Whyte, *More than a Legend,* 64
July 30, D. Maciver: *Inverness Courier,* 16 Jan. 1948
Aug.: Witchell, *Loch Ness Story,* 1976, p. 106
?, D. McBryde: Whyte, *More than a Legend,* 30
?, Arthur Grant: Costello, *In Search of Lake Monsters,* 77

1948

late May, Mr. Pearson: *Inverness Courier,* 15 June 1948
June 14: obs. no. 142 in Mackal, *Monsters of Loch Ness;* Whyte, *More than a Legend,* 193
July 15: Shields, *They Saw Nessie,* 22
Dec., Mrs. Ellice: Costello, *In Search of Lake Monsters,* 77; obs. no. 143 in Mackal, *Monsters of Loch Ness;* Whyte, *More than a Legend,* 39

1949

obs. no. 144 in Mackal, *Monsters of Loch Ness;* Whyte, *More than a Legend,* 50

1950

Apr. 19, Lady Baillie: obs. no. 145 in Mackal, *Monsters of Loch Ness;* Whyte, *More than a Legend,* 64; Witchell, *Loch Ness Story,* 1976, p. 93
June 20, C. Dunton: obs. no. 146 in Mackal, *Monsters of Loch Ness;* Shields, *They Saw Nessie,* 23; Whyte, *More than a Legend,* 60
Aug. 29, Miss McI.: obs. nos. 147–48 in Mackal, *Monsters of Loch Ness;* Shields, *They Saw Nessie,* 24; Whyte, *More than a Legend,* 193
Aug.?: Binns, *Loch Ness Mystery Solved,* 197

1951

Feb. 24: *Inverness Courier,* 27 Feb. 1951

June 27: ibid., 29 June 1951

June 28, R.B., R.McL.: obs. no. 149 in Mackal, *Monsters of Loch Ness;* Whyte, *More than a Legend,* 193

June ?, J. Harper Smith: Witchell, *Loch Ness Story,* 1976, p. 112

*July 14, Lachlan Stuart: ibid., 108; Burton, *Elusive Monster,* 74; photo 6 in Mackal, *Monsters of Loch Ness*

Nov. 13, P. Grant: Dinsdale, *Loch Ness Monster,* 98; obs. no. 162 in Mackal, *Monsters of Loch Ness;* Whyte, *More than a Legend,* 53; Witchell, *Loch Ness Story,* 1976, p. 112

1952

Feb. ?, D. Macdonald: *Inverness Courier,* 29 Feb. 1952

Apr. 1, P. MacMillan: Whyte, *More than a Legend,* 191

Apr. 28, Wm. MacDonald: ibid., 192; Burton, *Elusive Monster,* 137; Carruth, *Loch Ness and Its Monsters,* 19; obs. no. 151 in Mackal, *Monsters of Loch Ness;* Shields, *They Saw Nessie,* 25

July, McAffee: Costello, *In Search of Lake Monsters,* 79; Witchell, *Loch Ness Story,* 1976, p. 197

Aug. 18, Mrs. Finlay: Burton, *Elusive Monster,* 130; obs. no. 150 in Mackal, *Monsters of Loch Ness*

Aug. 20: *Inverness Courier,* 22 Aug. 1952

end Aug., Mr. Watkinson, *Daily Mail,* 26 Aug. 1952

Sept. 11: *Inverness Courier,* 12 Sept. 1952

Sept. 19, Neil MacCallum: *Evening Sentinel,* 2 Sept. 1983

Sept. 27, Vera McDonald: ibid.

Sept. ?, K. A. Key: Tim Dinsdale, *Project Water Horse,* London: Routledge and Kegan Paul, 1975, p. 187

?: *Nessletter,* no. 23.

?, Mr. Skinner: F. W. Holiday, *The Great Orm of Loch Ness,* London: Faber and Faber, 1968, p. 151 (all subsequent references are to this edition)

1953

Dec. 10, Tulloch: Burton, *Elusive Monster,* 126; obs. no. 152 in Mackal, *Monsters of Loch Ness;* Shields, *They Saw Nessie,* 26; Whyte, *More than a Legend,* 192

Dec. 15: *Daily Mail,* 16 Dec. 1953

1954

Cottier: Burton, *Elusive Monster,* 112

Feb. 4, David Slorach: Costello, *In Search of Lake Monsters,* 81; Hall, *Bestien, Scheusale und Monster,* 76

Feb. 26, A. J. Campbell: obs. no. 153 in Mackal, *Monsters of Loch Ness;* Shields, *They Saw Nessie,* 26; Whyte, *More than a Legend,* 192

July 9, Miss McDonald: *Inverness Courier,* 13 July 1954; obs. no. 154 in Mackal, *Monsters of Loch Ness;* Whyte, *More than a Legend,* 192
July 20, Miss E. McG.: obs. no. 156 in Mackal, *Monsters of Loch Ness;* Whyte, *More than a Legend,* 192
July ?, W. H. Davidson, Cary: Dinsdale, *Leviathans,* 196; obs. no. 155 in Mackal, *Monsters of Loch Ness;* Witchell, *Loch Ness Story,* 1976, p. 205
Aug. 8, Miss Ch. McInnes: *Inverness Courier,* 10 Aug. 1954; obs. no. 157 in Mackal, *Monsters of Loch Ness;* Whyte, *More than a Legend,* 191
Aug. 14, P. McMillan: *Inverness Courier,* 17 Aug. 1954; obs. no. 158 in Mackal, *Monsters of Loch Ness;* Whyte, *More than a Legend,* 191; Witchell, *Loch Ness Story,* 1976, p. 96
Sept. 9, S.M.: *Inverness Courier,* 17 Sept. 1954
Autumn, Miss Fraser: Whyte, *More than a Legend,* 172
Oct. 8: obs. no. 159 in Mackal, *Monsters of Loch Ness;* Whyte, *More than a Legend,* 43; Witchell, *Loch Ness Story,* 1976, p. 95
?, Mrs. McNaughton: *Inverness Courier,* 11 Sept. 1964

1955

Scots Magazine, Oct. 1975
Apr., R. Ward: obs. no. 160 in Mackal, *Monsters of Loch Ness;* Whyte, *More than a Legend,* 38
early July, Miss Fraser: obs. no. 161 in Mackal, *Monsters of Loch Ness;* Whyte, *More than a Legend,* 43
July 27, Mr. A. Mackenzie: *Inverness Courier,* 29 July 1955
*July 29, P. A. Macnab: Costello, *In Search of Lake Monsters,* 82; photo 7 in Mackal, *Monsters of Loch Ness;* Witchell, *Loch Ness Story,* 1976, p. 127
early Aug., K. Shakespeare: Burton, *Elusive Monster,* 117; Tim Dinsdale, *The Story of the Loch Ness Monster,* London: Target, 1973, p. 68

1955/56

Alex Campbell: Binns, *Loch Ness Mystery Solved,* 79; Witchell, *Loch Ness Story,* 1976, p. 82

1956

July, Mr. & Mrs. Graham: obs. no. 163 in Mackal, *Monsters of Loch Ness*

1957

Mar. 11, Mr. D. Fowlie: *Inverness Courier,* 18 June 1957
Mar. 11, J. Grant: Shields, *They Saw Nessie,* 28; Witchell, *Loch Ness Story,* 1976, p. 118
June 16, D. Campbell: Dinsdale, *Loch Ness Monster,* 115; obs. no. 164 in Mackal, *Monsters of Loch Ness;* Witchell, *Loch Ness Story,* 1976, p. 96

June 16, Mr. & Mrs. McQueen: *Inverness Courier,* 18 June 1957
July, Mr. Jan Verleun: *Daily Express,* 25 July 1957
Dec. 29, R. Bain: Shields, *They Saw Nessie,* 29; Witchell, *Loch Ness Story,* 1976, p. 120

1958

Spring, Mr. & Mrs. Rowland: Burton, *Elusive Monster,* 77; obs. no. 165 in Mackal, *Monsters of Loch Ness*
May 16, William Grigor: *Press and Journal,* 22 May 1958
May 21: Shields, *They Saw Nessie,* 29
May ?, Watchorn: Dinsdale, *Leviathans,* 200
July 16, Alex Campbell: Binns, *Loch Ness Mystery Solved,* 79; Witchell, *Loch Ness Story,* 1976, p. 81
July 16: Shields, *They Saw Nessie,* 30
Aug. 5, M. Scott Moncrieff: *Highland Herald,* 7 Aug. 1958
*Oct., H. L. Cockrell: photo 8 in Mackal, *Monsters of Loch Ness;* Witchell, *Loch Ness Story,* 1976, p. 125
Oct., Brown: Dinsdale, *Loch Ness Monster,* 102

1959

Feb. 2, H. McIntosh: Shields, *They Saw Nessie,* 31; Witchell, *Loch Ness Story,* 1976, p. 97
Mar. 22, Denys Tucker: Costello, *In Search of Lake Monsters,* 83
late June, A. Andrews: *Nessletter,* no. 5
July 4: *Inverness Courier,* 10 July 1959
July 5: ibid.
July 8: Shields, *They Saw Nessie,* 32
July 10, R. Murray: Dinsdale, *Loch Ness Monster,* 58; Shields, *They Saw Nessie,* 31
July 20: *Times,* 20 July 1959
Aug./Sept., R. F. Sisson: Alan Villiers, "Scotland from Her Lovely Lochs and Seas," *National Geographic,* Apr. 1961, pp. 492–539

1960

Feb. 28, T. McLeod: land sight. no. 17 in Mackal, *Monsters of Loch Ness;* Witchell, *Loch Ness Story,* 1976, p. 141
Apr. 20: Dinsdale, *Loch Ness Monster,* 71
*Apr. 23, Tim Dinsdale: Dinsdale, *Loch Ness Monster,* 76; film 7 in Mackal, *Monsters of Loch Ness;* Witchell, *Loch Ness Story,* 1976, p. 147
May 24, P. O'Connor: obs. no. 166 in Mackal, *Monsters of Loch Ness*
*May 27, P. O'Connor: photo 9 in ibid.; Burton, *Elusive Monster,* 81; Costello, *In Search of Lake Monsters,* 87
June 20, M. Burton: Burton, *Elusive Monster,* 170
June 23, Jane Burton: Maurice Burton, "The Problem of the Loch Ness Monster: A Scientific Investigation," *Illustrated London News,* 1960

*June 24, M. Burton: Burton, *Elusive Monster,* 161; Costello, *In Search of Lake Monsters,* 88; photo 10 in Mackal, *Monsters of Loch Ness*
June 26: Burton, "Problem of the Loch Ness Monster"
July 3, P. O'Connor: obs. no. 168 in Mackal, *Monsters of Loch Ness*
July 4: obs. no. 167 in ibid.; Burton, *Elusive Monster,* 66; Costello, *In Search of Lake Monsters,* 89
July 10, Bruce Ing: Burton, *Elusive Monster,* 65; Costello, *In Search of Lake Monsters,* 89; obs. no. 169 in Mackal, *Monsters of Loch Ness;* Witchell, *Loch Ness Story,* 1976, p. 150
July 11: Shields, *They Saw Nessie,* 32
July 14, Bruce Ing: obs. no. 170 in Mackal, *Monsters of Loch Ness*
July ?, J. Edwards: Burton, *Elusive Monster,* 111
*Aug. 7, McLeods: obs. no. 171 in Mackal, *Monsters of Loch Ness;* Witchell, *Loch Ness Story,* 1976, p. 151
Aug. 13, Dobb: obs. no. 172 in Mackal, *Monsters of Loch Ness*
Aug. ?: Burton, *Elusive Monster,* 118
Aug. ?, Mrs. Cary: Dinsdale, *Leviathans,* 196
Oct. 23, Rev. Dobb: Burton, *Elusive Monster,* 11; obs. no. 173 in Mackal, *Monsters of Loch Ness*
Nov. 27: Shields, *They Saw Nessie,* 33
Dec. 5: ibid., 34; Burton, *Elusive Monster,* 26; obs. no. 174 in Mackal, *Monsters of Loch Ness*
Dec. ?, Robert Duff: obs. no. 175 in Mackal, *Monsters of Loch Ness*

1961

June: *The Field,* 23 Nov. 1961
June 8, David Foster: Shields, *They Saw Nessie,* 35
July 21, H. Stiff: Dinsdale, *Leviathans,* 64
Aug. 11, Stanley Hill: Costello, *In Search of Lake Monsters,* 90
Aug. 11, Bert McDonald: *Field,* 23 Nov. 1961
Aug. ?, McIntosh: obs. no. 176 in Mackal, *Monsters of Loch Ness*
Nov.: Holiday, *Great Orm,* 9

1962

G.A.S.: Dinsdale, *Leviathans,* 39
May 11, Mrs. Christie: ibid., 30; Costello, *In Search of Lake Monsters,* 91
June 29, H. G. Hasler: Costello, *In Search of Lake Monsters,* 94
June ?: ibid., 95; Mackal, *Monsters of Loch Ness,* 351
Aug. 23, F. W. Holiday: Holiday, *Great Orm,* 11
Aug. 23: ibid., 10
Aug. 23, F. W. Holiday: ibid., 13
Aug. 24, F. W. Holiday: Hall, *Bestien, Scheusale und Monster,* 78; obs. no. 177 in Mackal, *Monsters of Loch Ness*
Aug. ?, Arthur Kopit: *New York Times Magazine,* 1 Aug. 1976

*Oct. 19: Costello, *In Search of Lake Monsters,* 97; Hall, *Bestien, Scheusale und Monster,* 78; film 8 in Mackal, *Monsters of Loch Ness;* Witchell, *Loch Ness Story,* 1976, p. 156
Oct. 19, Michael Spear: Costello, *In Search of Lake Monsters,* 97; Witchell, *Loch Ness Story,* 1976, p. 156
Oct. 25: Costello, *In Search of Lake Monsters,* 97

1963

Apr. 21, Mrs. Vera King: *Sun World Telegram,* 22 Apr. 1963
June 4: *Inverness Courier,* 20 Mar. 1964
June 4, Sylvia MacKintosh: obs. no. 178 in Mackal, *Monsters of Loch Ness*
June 5: *Inverness Courier,* 20 Mar. 1964
June 6: ibid.
*June 6: film 10, land sight. no. 18 in Mackal, *Monsters of Loch Ness;* Witchell, *Loch Ness Story,* 1976, p. 144
*June 6: film 9 in Mackal, *Monsters of Loch Ness*
June 12, Jack MacLean: Shields, *They Saw Nessie,* 36
*June 13: *Inverness Courier,* 11 Oct. 1963; film 11 in Mackal, *Monsters of Loch Ness*
July 31, John MacLean: *Inverness Courier,* 2 Aug. 1963
July ?, D. MacIntosh: Dinsdale, *Leviathans,* 15; Witchell, *Loch Ness Story,* 1976, p. 179
Aug., F. W. Holiday: Holiday, *Great Orm,* 43
Aug. 20, R. West: *Inverness Courier,* 23 Aug. 1963; Shields, *They Saw Nessie,* 36
Aug. 21, E. Kerr: *Inverness Courier,* 23 Aug. 1963
Aug. 23, W. Fallows: ibid., 27 Aug. 1963
Aug. 24, D. Blacklock: ibid.
Aug. ?, Heather Cary: Dinsdale, *Leviathans,* 196
Aug. ?, A. Grant: obs. nos. 179–80 in Mackal, *Monsters of Loch Ness;* Witchell, *Loch Ness Story,* 1976, p. 97
?, Doglerush: *Nessletter,* no. 18

1964

Jan. 11, Alex Campbell: obs. no. 181 in Mackal, *Monsters of Loch Ness;* Dinsdale, *Leviathans,* 205
Mar. 25, A. Russel: Dinsdale, *Leviathans,* 205; obs. no. 183 in Mackal, *Monsters of Loch Ness;* Shields, *They Saw Nessie,* 37
Mar. ?, William Fraser: obs. no. 182 in Mackal, *Monsters of Loch Ness*
May 6: obs. no. 184 in ibid.; Dinsdale, *Leviathans,* 205
May 17, Mr. & Mrs. Eames: Dinsdale, *Leviathans,* 205; obs. no. 185 in Mackal, *Monsters of Loch Ness*
May 18, Fred Pullen: Dinsdale, *Leviathans,* 205; obs. no. 186 in Mackal, *Monsters of Loch Ness*
May 19, Mr. & Mrs. Hodge, Ivor Newby: Costello, *In Search of Lake*

Monsters, 100; Dinsdale, *Leviathans,* 205; obs. no. 187 in Mackal, *Monsters of Loch Ness;* Witchell, *Loch Ness Story,* 1976, pp. 161–62
May 27: Witchell, *Loch Ness Story,* 1976, p. 178
June 9: Dinsdale, *Leviathans,* 205
July 17, Alex Campbell: ibid., 206; obs. no. 188 in Mackal, *Monsters of Loch Ness*
Aug. 9, J. Dawson: Dinsdale, *Leviathans,* 206; obs. no. 189 in Mackal, *Monsters of Loch Ness*
Aug. 19, J. Thompson: Dinsdale, *Leviathans,* 206; obs. no. 190 in Mackal, *Monsters of Loch Ness*
Aug. 30: Dinsdale, *Leviathans,* 206; obs. no. 191 in Mackal, *Monsters of Loch Ness*
Sept. 5: ibid.
Sept. 6, Mr. & Mrs. Riley: Dinsdale, *Leviathans,* 206; obs. no. 192 in Mackal, *Monsters of Loch Ness*
Sept. 8, Mrs. McNaughton: Dinsdale, *Leviathans,* 206; *Inverness Courier,* 11 Sept. 1964; obs. no. 193 in Mackal, *Monsters of Loch Ness*
early Oct., Mrs. Dalles: Dinsdale, *Leviathans,* 206; obs. no. 194 in Mackal, *Monsters of Loch Ness*
mid-Oct.: Dinsdale, *Leviathans,* 206; obs. no. 195 in Mackal, *Monsters of Loch Ness*

1965

Mar. 30, Miss Keith: obs. no. 196 in Mackal, *Monsters of Loch Ness;* Witchell, *Loch Ness Story,* 1976, p. 180
Apr. 14, W. J. Home: Holiday, *Great Orm,* 157
June, Frank Searle: Hall, *Bestien, Scheusale und Monster,* 79
June: obs. no. 197 in Mackal, *Monsters of Loch Ness*
June, F.W. Holiday: obs. no. 198 in ibid.; Costello, *In Search of Lake Monsters,* 101
June 6, Simon Campbell: Holiday, *Great Orm,* 100
June 15, Mr. Cameron & Mr. Fraser: ibid., 107
June 21, F. W. Holiday: Costello, *In Search of Lake Monsters,* 102; obs. no. 199 in Mackal, *Monsters of Loch Ness*
July 30, H. Ferguson: obs. no. 200 in Mackal, *Monsters of Loch Ness*
*Aug. 1, E. Hall: Costello, *In Search of Lake Monsters,* 102; film 13 in Mackal, *Monsters of Loch Ness*
Sept. 8, Helen McNaughton: Holiday, *Great Orm,* 61
Sept. 30, E. & V. Eliot: obs. no. 201 in Mackal, *Monsters of Loch Ness*
?, Br. Carruth: Witchell, *Loch Ness Story,* 1976, p. 102

1966

Mar. 27, David James: Costello, *In Search of Lake Monsters,* 104
Apr. 21, Hilda Hitching: Shields, *They Saw Nessie,* 38
May 28, Mr. & Mrs. Macdonald: obs. no. 202 in Mackal, *Monsters of Loch Ness*

May 29, Mr. & Mrs. Pommitz: obs. no. 203 in ibid.
May 31, Mae Macdonald: obs. no. 204 in ibid.
June 13, B. Cameron: obs. no. 205 in ibid.
June 14, F. Young: obs. no. 206 in ibid.
June 14, Mr. & Mrs. Johnstone: obs. no. 207 in ibid.
June 20, L. Holmgren: obs. no 208 in ibid.
June 29, R. W. Swan: obs. no. 209 in ibid.
July 7, Mr. & Mrs. Dickson: obs. no. 210 in ibid.
July 21, Clem Skelton: obs. no. 211 in ibid.
July 28, H. Cary: obs. no. 212 in ibid.
July 28, Knapp: obs. no. 213 in ibid.
July 30, Miss Lewis: obs. no. 214 in ibid.
Aug. 20, Dr. Heathcote: obs. nos. 215–16 in ibid.
*Aug. 20, Sandermann: photo 13 in ibid.
Sept. 5, M. Pool: obs. no. 217 in ibid.
Sept. 17, Roy P. Mackal: ibid., 20
Sept. 25, Clem Skelton, Peter Hodge, Angela Veitch, Roy Mackal: obs. no. 218 in ibid., and pp. 20–21
Sept. 28, Lichterfelde: Witchell, *Loch Ness Story,* 1976, p. 99
Sept. 28, Mrs. N. Shulman: obs. no. 219 in Mackal, *Monsters of Loch Ness*
?, Roy P. Mackal: ibid., 19
*?, M. Edwards: Witchell, *Loch Ness Story,* 1976, p. 164
?, F. Gregory: obs. no. 210 in Mackal, *Monsters of Loch Ness*

1967

Mar. 17, John Cameron: obs. no. 221 in ibid.; David James, *Loch Ness Investigation Annual Report,* n.p., 1967, p. 3; Witchell, *Loch Ness Story,* 1976, p. 101
Apr. 5, D. Wathen: James, *LNI Annual Report,* 1967, p. 3; obs. no. 222 in Mackal, *Monsters of Loch Ness*
Apr. 14, Mr. & Mrs. Cary: James, *LNI Annual Report,* 1967, p. 3; obs. no. 223 in Mackal, *Monsters of Loch Ness*
Apr. ?, D. Fraser: Witchell, *Loch Ness Story,* 1976, p. 99
*May 22, L. S. Durkin, Miss Atkin: James, *LNI Annual Report,* 1967, p. 4; film 14, obs. no. 224 in Mackal, *Monsters of Loch Ness;* Witchell, *Loch Ness Story,* 1976, p. 165
May 27, F. W. Holiday: James, *LNI Annual Report,* 1967, p. 4
May 28, Pyman: ibid.; obs. no. 225 in Mackal, *Monsters of Loch Ness*
June 7, R. Dodson: James, *LNI Annual Report,* 1967, p. 4
*June 13, R. H. Raynor: ibid.; film 15, obs. no. 226 in Mackal, *Monsters of Loch Ness;* Witchell, *Loch Ness Story,* 1976, p. 165
*July 15, Dobbie: James, *LNI Annual Report,* 1967 p. 5; photo 14, obs. no. 227 in Mackal, *Monsters of Loch Ness*
July 23: Costello, *In Search of Lake Monsters,* 117

Aug. 6, Peter Davies: James, *LNI Annual Report,* 1967, p. 5; obs. no. 228 in Mackal, *Monsters of Loch Ness*
Aug. 7, N. Schofield: James, *LNI Annual Report,* 1967, p. 5; obs. no. 229 in Mackal, *Monsters of Loch Ness*
Aug. 13, J. Kendrick: James, *LNI Annual Report,* 1967, p. 5
Aug. 18, D. Field: ibid.; obs. no. 230 in Mackal, *Monsters of Loch Ness*
Aug. 22, D. Gartrell: James, *LNI Annual Report,* 1967, p. 6; obs. no. 231 in Mackal, *Monsters of Loch Ness*
*Aug. 22, Andrew Chapman: James, *LNI Annual Report,* 1967, p. 6; film 16 in Mackal, *Monsters of Loch Ness*
*Aug. 23, S. Hunter: James, *LNI Annual Report,* 1967, p. 6; film 17 in Mackal, *Monsters of Loch Ness*
Aug. 29, Kewley: James, *LNI Annual Report,* 1967, p. 7; obs. no. 232 in Mackal, *Monsters of Loch Ness*
*Summer: Costello, *In Search of Lake Monsters,* 117
Sept. 18, John Matheson: James, *LNI Annual Report,* 1967, p. 7; Mackal, *Monsters of Loch Ness,* 28
Sept. 20, D. Bland: James, *LNI Annual Report,* 1967, p. 7; obs. no. 233 in Mackal, *Monsters of Loch Ness*
Sept. 21, A. Hayward: James, *LNI Annual Report,* 1967, p. 8; obs. no. 234 in Mackal, *Monsters of Loch Ness*
Sept. 26, John Stout: James, *LNI Annual Report,* 1967, p. 8; obs. no. 235 in Mackal, *Monsters of Loch Ness*
*Oct. 5, Clem Skelton: James, *LNI Annual Report,* 1967, p. 8; film 18 in Mackal, *Monsters of Loch Ness*

1968

Mr. W. A. Adamson: Holiday, *Great Orm,* 41
Apr. 18, C. Sanders: David James, *Loch Ness Investigation Annual Report,* n.p., 1968, p. 2; obs. no. 236 in Mackal, *Monsters of Loch Ness*
Apr. 18, W. V. Turl: James, *LNI Annual Report,* 1968, p. 2; obs. no. 237 in Mackal, *Monsters of Loch Ness*
May 3, G. Brusey: Shields, *They Saw Nessie,* 40
May 4, L. Irvine: James, *LNI Annual Report,* 1968, p. 2; obs. no. 238 in Mackal, *Monsters of Loch Ness*
May 5, P. Bull: James, *LNI Annual Report,* 1968, p. 2; obs. no. 239 in Mackal, *Monsters of Loch Ness*
May 10, Mrs. Wolbourn: James, *LNI Annual Report,* 1968, p. 3
May 27, Mr. & Mrs. Warren: obs. no. 240 in Mackal, *Monsters of Loch Ness*
June 27, Lundberg: obs. no. 241 in ibid.; James, *LNI Annual Report,* 1968, p. 3
July 7, John Mackay: Shields, *They Saw Nessie,* 41
July 10, Mr. & Mrs. Deacon: James, *LNI Annual Report,* 1968, p. 3; obs. no. 242 in Mackal, *Monsters of Loch Ness*

July 23, Mr. & Mrs. Heal: James, *LNI Annual Report,* 1968, p. 3; obs. no. 243 in Mackal, *Monsters of Loch Ness*
July 27, McLean: James, *LNI Annual Report,* 1968, p. 3
July ?, Dr. K. McLeod: Bord & Bord, *Alien Animals,* 38
Aug. 23, Hugh Munro: Shields, *They Saw Nessie,* 41
Aug. 26, F. W. Holiday: James, *LNI Annual Report,* 1968, p. 3; obs. no. 244 in Mackal, *Monsters of Loch Ness*
Aug. 26, Thresh: James, *LNI Annual Report,* 1968, p. 3
Sept. 4, Grummet: ibid.; obs. no. 245 in Mackal, *Monsters of Loch Ness*
Sept. 5, Ian Smith: Dinsdale, *Project Water Horse,* 62
Sept. 5: ibid., 64
Sept. 6, Dinsdale: ibid.; Dinsdale, *Loch Ness Monster,* 128
Sept. 19, Mr. & Mrs. Silcock: James, *LNI Annual Report,* 1968, p. 3; obs. no. 246 in Mackal, *Monsters of Loch Ness*
Nov. 11, McLeod: James, *LNI Annual Report,* 1968, p. 4; obs. no. 247 in Mackal, *Monsters of Loch Ness*

1969

Apr. 7, B. Marshall: David James, *Loch Ness Investigation Annual Report,* n.p., 1969, p. 5; obs. no. 248 in Mackal, *Monsters of Loch Ness*
*May 27, H. Barski: James, *LNI Annual Report,* 1969, p. 5; film 20 in Mackal, *Monsters of Loch Ness*
June 1, W. Cary: James, *LNI Annual Report,* 1969, p. 5; Witchell, *Loch Ness Story,* 1976, p. 206; Witchell, *Loch Ness Story,* 1982, p. 164
June 5, Best: James, *LNI Annual Report,* 1969, p. 5
June 21, Mrs. Matthews: ibid.
*June 23, A. Skelton: ibid.; film 21 in Mackal, *Monsters of Loch Ness*
July 26, Mr. & Mrs. Clayton: James, *LNI Annual Report,* 1969, p. 5; obs. no. 249 in Mackal, *Monsters of Loch Ness*
Aug. 1, R. Moyse: James, *LNI Annual Report,* 1969, p. 6; obs. no. 250 in Mackal, *Monsters of Loch Ness*
Aug. 6, Craven: James, *LNI Annual Report,* 1969, p. 6; obs. no. 251 in Mackal, *Monsters of Loch Ness*
Aug. 6, Graham Bayley: James, *LNI Annual Report,* 1969, p. 6
Aug. 12, James Skeldon: ibid.; *Nessletter,* no. 55
*Aug. ?, Mrs. Tait: Hall, *Bestien, Scheusale und Monster,* 64; Costello, *In Search of Lake Monsters,* 112
*Sept. 16, I. Shield: James, *LNI Annual Report,* 1969, p. 6; film 22 in Mackal, *Monsters of Loch Ness*
Sept. ?: Dinsdale, *Project Water Horse,* 189
Oct. 10, M. Rickards: James, *LNI Annual Report,* 1969, p. 7

1970

July 18, Mr. & Mrs. Tyrrel: Witchell, *Loch Ness Story,* 1976, p. 180
July 26: Costello, *In Search of Lake Monsters,* 111
Aug., Tim Dinsdale: Binns, *Loch Ness Mystery Solved,* 142; Carruth, *Loch Ness and Its Monsters,* 23; Dinsdale, *Loch Ness Monster,* 207

Sept., Jack Uhlrich: Costello, *In Search of Lake Monsters,* 111
Sept., Roy P. Mackal: Mackal, *Monsters of Loch Ness,* 69

1971

May, S. France: Witchell, *Loch Ness Story,* 1976, p. 181
June 22, William Dewar: ibid., 102
*June 23, Bob Rines, Basil Cary: ibid., 185
early Aug., B. Badger: Dinsdale, *Loch Ness Monster,* 144
Aug. 4, F. Searle: *Frankfurter Allgemeine Zeitschrift,* 30 Oct. 1976
Aug. 18, Katherine Robertson: Dinsdale, *Story of the Loch Ness Monster,* 82
Sept. 6, Tim Dinsdale: Dinsdale, *Loch Ness Monster,* 146
Oct. 13, Mrs. Turner: ibid., 148
Oct. 13, H. Henderson: ibid., 150; Witchell, *Loch Ness Story,* 1976, p. 102
Oct. 14, Gregory Brusey: Witchell, *Loch Ness Story,* 1976, p. 102
*Nov. 10, F. Searle: *Frankfurter Allgemeine Zeitschrift,* 30 Oct. 1976

1972

June, F. Searle: ibid.
*July 27, F. Searle: photo 15 in Mackal, *Monsters of Loch Ness;* Witchell, *Loch Ness Story,* 1976, p. 185
early Aug., E. Essex: Dinsdale, *Story of the Loch Ness Monster,* 69
Aug. 1, Nick Witchell: Binns, *Loch Ness Mystery Solved,* 166
*Aug. 8, Bob Rines: photo 16 in Mackal, *Monsters of Loch Ness*
*Oct. 21, F. Searle: *Frankfurter Allgemeine Zeitschrift,* 30 Oct. 1976; Witchell, *Loch Ness Story,* 1976, p. 184
?, Comdr. Bellars: Dinsdale, *Project Water Horse,* 203

1973

*Mar. 27, F. Searle: Witchell, *Loch Ness Story,* 1976, p. 185
Spring: *Nessletter,* no. 1
May, Ian McKenzie: ibid., no. 3
June: Dinsdale, *Project Water Horse,* 190
July 27, J. Shaw: *Nessletter,* no. 1; Witchell, *Loch Ness Story,* 1976, pp. 101, 103
July 27, Pugh: *Nessletter,* no. 2
*Aug. 1, F. Searle: Witchell, *Loch Ness Story,* 1976, p. 185
Aug. 5, Nick Witchell: *Nessletter,* no. 1
Aug. 20, Graham Snape: ibid.
Sept. 8/9, Ian Henderson: ibid.
Nov. 10, R. Jenkyns: ibid., no. 3; Mackal, *Monsters of Loch Ness,* 265; Shields, *They Saw Nessie,* 41; Witchell, *Loch Ness Story,* 1976, p. 200
?, Mme. Clerc: Jean-Jacques Barloy, *Newsletter,* published in Paris, no. 10

1974

*Jan. 8, F. Searle: Hall, *Bestien, Scheusale und Monster,* 83; Witchell, *Loch Ness Story,* 1976, p. 185

Feb. 8, A. Call: N. Blundell, *The Greatest Mysteries,* London: Octopus, 1980, p. 68; *Nessletter,* no. 2
Apr.: *Nessletter,* no. 3
May, Ian McKenzie: ibid.
June 20, G. Siegel: ibid., no. 4
June 22, D. Steward: ibid.
July 5, D. Neal: ibid.
*July 9, F. Searle: *Frankfurter Allgemeine Zeitschrift,* 30 Oct. 1976
July ?, L. J. Prince: *Nessletter,* no. 26
Sept. 30, Mr. & Mrs. Jenkyns: ibid., no. 6; Bord & Bord, *Alien Animals,* 38, 42; Dinsdale, *Project Water Horse,* 204
Oct. 6, John Keay: *Nessletter,* no. 5; Shields, *They Saw Nessie,* 42
early Nov., K. Buggy: *Nessletter,* nos. 6–7

1975

Apr. 21, J. Berton: ibid., no. 9
*June 20, R. Rines: photo 17 in Mackal, *Monsters of Loch Ness;* Witchell, *Loch Ness Story,* 1976, p. 207
July 7: *Nessletter,* no. 10; Shields, *They Saw Nessie,* 43
July 10, Sue Diamond: *Nessletter,* nos. 10–11
*July 18, Alan Wilkins: ibid., no. 10
July ?, Tim Dinsdale: Dinsdale, *Loch Ness Monster,* 163
Aug. 4: *Nessletter,* no. 10
Oct., R. Lipinski: ibid., no. 11
?: Blundell, *Greatest Mysteries,* 68

1976

William S. Ellis, "Loch Ness—the Lake and the Legend," *National Geographic,* June 1977, pp. 758–79
early: *Weltwoche* (Zurich), 23 June 1976
Jan. 10, F. Searle: Frank Searle, *Quarterly Newsletter,* Loch Ness Investigation, Lower Foyers, Inverness, Apr. 1977
*Feb. 26, F. Searle: ibid.
Feb. 28, Rainford: ibid.
Mar. 19, Downie: ibid.
Apr. 13, D. Nicholson: *Inverness Courier,* 16 Apr. 1976; *Nessletter,* no. 15
May 4, Mr. & Mrs. Ely: Searle, *Quarterly Newsletter,* Apr. 1977
May 4, Tom Skinner: ibid.
May ?, Roger Parker: Witchell, *Loch Ness Story,* 1982, p. 200
June 7, Tolchard: Searle, *Quarterly Newsletter,* Apr. 1977
July 1, A. McKenzie: ibid.
*July 4, Alan Landsburg: Alan Landsburg, *In Search of Myths and Monsters,* New York: Bantam, 1977, p. 69
mid-year, B. Kennedy: Ellis, "Lake and the Legend," 758–79; *Nessletter,* no. 17

Aug. 1, Nick Asby: *Nessletter,* no. 17
Aug. 21, Anthony Luke: ibid.; *Inverness Courier,* 24 Aug. 1976; Searle, *Quarterly Newsletter,* Apr. 1977
*Aug. 22; F. Searle: Searle, *Quarterly Newsletter,* Apr. 1977
*Aug. 25, Rev. MacNaughton: *Nessletter,* no. 18
*Aug. ?, B. & L. Green: ibid.
Aug. ?, Ph. Marlow: ibid., no. 20
Sept. 7, Sampson: ibid., nos. 18, 21
Oct. 11, Brian Cosens: ibid., no. 18; *Inverness Courier,* 15 Oct. 1976
Nov. 6, F. Searle: Searle, *Quarterly Newsletter,* Apr. 1977
Dec. 2, Mrs. McIntosh: ibid.

1977

Jan. 4, J. Bradley: ibid.
Jan. 22, Sheila Ward: ibid.
Jan. 31-Feb. 2: Bord & Bord, *Alien Animals,* 195
Mar. 24, Mr. & Mrs. Watkins: Searle, *Quarterly Newsletter,* Apr. 1977
Spring: Barloy, *Newsletter,* no. 1
Apr. 27, McGrew: *Nessletter,* no. 22; Searle, *Quarterly Newsletter,* Apr. 1977
Apr. 30, Mr. & Mrs. Walne: Searle, *Quarterly Newsletter,* June 1977
*May 21, Mr. & Mrs. T. Shields: Bord & Bord, *Alien Animals,* 26–27; Dinsdale, *Loch Ness Monster,* 184
June 21, Fallon: Searle, *Quarterly Newsletter,* Sept. 1977
Aug. 8, F. Searle: ibid.
*Aug. 22, Peter Smith: Dinsdale, *Loch Ness Monster,* 190; *Nessletter,* no. 23; Witchell, *Loch Ness Story,* 1982, p. 204
early Sept., P. McFarlane: Dinsdale, *Loch Ness Monster,* 190; *Nessletter,* no. 36
Sept. 8, Mrs. Perry: Searle, *Quarterly Newsletter,* Dec. 1977
Sept. 15, C. E. Moore: ibid.
Oct. 29, J. Murray: *Nessletter,* no. 25
Nov., Caroline: ibid., no. 30
?, H. Hepple: ibid., no. 26

1978

Mar. 6: Searle, *Quarterly Newsletter,* Mar. 1978
Apr. 19, J. Cameron: ibid., June 1978
early May, Susan Russell: *Nessletter,* no. 28
May 26, G. Hall: Searle, *Quarterly Newsletter,* June 1978
June 9, F. Searle: ibid.
June 15, A. McMillan: ibid.
June 17, Bill Wright: ibid., Sept. 1978; Bord & Bord, *Alien Animals,* 23; *Nessletter,* no. 28; Shields, *They Saw Nessie,* 43
July 9, Mrs. Gibson: Searle, *Quarterly Newsletter,* Sept. 1978
July 24, Miss Page: *Nessletter,* no. 15

July 25, Painten: Searle, *Quarterly Newsletter,* Sept. 1978
July 27, Mark Lawson: ibid.
Aug. 3, P. Locas: ibid.
Aug. 5, Mr. & Mrs. Chisholm: Blundell, *Greatest Mysteries,* 68; *Nessletter,* no. 29; Shields, *They Saw Nessie,* 44
Aug. 8, Richardson: Searle, *Quarterly Newsletter,* Sept. 1978
*Sept. 3, Jeoff Watson: Dinsdale, *Loch Ness Monster,* 175
Sept. 7, Mrs. Rorvik: Searle, *Quarterly Newsletter,* Dec. 1978
Oct. 31, Mrs. Legdon: ibid.
Dec., F. Searle: ibid.
?: *Nessletter,* no. 28

1979

Lord Lovat: *Zweites Deutsches Fernsehen* (Mainz), 28 Sept. 1979
Jan. 3, J. Bennet: Searle, *Quarterly Newsletter,* Mar. 1979
Feb. 20, F. Searle: ibid.
Mar. 25: ibid., June 1979
Mar. 28, Hepworth: ibid.
*Apr. 2, B. Hamilton: *Daily Telegraph,* 1 Apr. 1972, p. 15
Apr. 5, R. Poyner: Searle, *Quarterly Newsletter,* June 1979
Apr. 18, Alec Welch: ibid.
May 8, G. Hamilton: ibid.
June 6, Ken Nash: ibid., Sept. 1979
June 12, C. Kennedy: Dinsdale, *Loch Ness Monster,* 178
July 7, J. Alderdice: Searle, *Quarterly Newsletter,* Sept 1979
July 14, G. Millar: ibid.
July 26, Mr. & Mrs. Ryan: ibid.
July 27: *Nessletter,* no. 36
July 30, Alastair Boyd: ibid., nos. 59–64
Aug. 6, R. Acraman: ibid., no. 36; Barloy, *Newsletter,* 13
Aug. 12: *Sunday Post* (Glasgow), 12 Aug. 1979
Aug. 22, F. Knight: Searle, *Quarterly Newsletter,* Sept. 1979
Aug. 24, Cresswell: ibid.
Aug. 31, Mrs. Mcleod: *Nessletter,* no. 36
Sept. 6, Mrs. McNish: ibid., nos. 36, 50
Sept. 9, A. Kennedy: ibid., no. 37
Sept. 21, Flint: Searle, *Quarterly Newsletter,* Dec. 1979
Sept. 26, Miss Pat: ibid.
Nov. 10, Mr. & Mrs. Kirkup: ibid.
?, Ray Greenfield: *Nessletter,* no. 53

1980

Jan. 25, Mr. & Mrs. Rhodes: Searle, *Quarterly Newsletter,* Mar. 1980
Feb. 3, Mr. & Mrs. Wilkinson: ibid.
Feb. 10, M. Hay: ibid.
Feb. 21, Walton: ibid.

Mar. 10, F. Searle: ibid., June 1980
Mar. 27, Catlow: ibid., Sept. 1980
Apr. 7, Quinn: ibid., June 1980
Apr. 23, Fowler: ibid.
May 29, F. Searle: ibid., Sept. 1980
June 17, Davies: ibid.
June 27, J. Wagstaff: ibid.
Aug. 2, Gibbs: ibid.
Aug. 8, G. Prideaux: ibid.
Aug. 23, Jim Green: *Nessletter,* no. 42
Sept. 13, Aston: Searle, *Quarterly Newsletter,* Dec. 1980
Nov. 16, F. Searle: ibid.
Dec. 8, J. Cameron: *Fortean Times,* no. 34; *Nessletter,* no. 43

1981

Jan. 11, Mr. & Mrs. Davidson: Searle, *Quarterly Newsletter,* Mar. 1981
Feb. 25, N. Wylie: ibid.
Apr. 4, P. & H. Johnston: ibid., June 1981
Apr. 14, Mr. & Mrs. Knight: ibid.
Apr. 27, Dutton: ibid.
May 15, Mr. & Mrs. Exley: ibid.
June 23, F. Searle: ibid., Sept. 1981
June 24, Mr. & Mrs. Cunningham: ibid.
July: *Nessletter,* nos. 50, 52
Aug. 6, Grimshaw: ibid., no. 64
Aug. 6, Mrs. Metcalfe: *Sunday Post,* 6 Sept. 1981
Aug. 9, Mr. & Mrs. Vargas: Searle, *Quarterly Newsletter,* Sept. 1981
Aug. 15, F. Searle: ibid.
late Aug.: *Nessletter,* no. 2
Sept. 10, Alice Bjornstad: ibid., no. 49
Sept. 21, Jill Donovan: Searle, *Quarterly Newsletter,* Dec. 1981
*Sept. ?: *Nessletter,* no. 52
Nov. 7, B. Dennison: Searle, *Quarterly Newsletter,* Dec. 1981
*?, David Bead: *Nessletter,* no. 50

1982

Jan. 23, Christie: Searle, *Quarterly Newsletter,* Mar. 1982
Feb. 27: ibid.; *Nessletter,* no. 52
Apr. 4, F. Searle: Searle, *Quarterly Newsletter,* June 1982
Apr. 30: ibid.; *Nessletter,* no. 52
May 20, Rosen: ibid.
May 23: *Nessletter,* nos. 52–53; Searle, *Quarterly Newsletter,* June 1982
July 10, Ken Armstrong: Searle, *Quarterly Newsletter,* Sept. 1982
July 16, G. Roby: ibid.; *Daily Record,* 16 July 1982
July 20, Mr. & Mrs. Urmston: Searle, *Quarterly Newsletter,* Sept. 1982
July 28: ibid.

late July, R. Acraman: *Nessletter,* no. 53
Aug. 2, James Fisher: ibid.; Shields, *They Saw Nessie,* 46
Aug. 14, Paul Marsdon: Searle, *Quarterly Newsletter,* Sept. 1982
Aug. 26, Stan Morris: ibid.
*Aug. ?, G. & J. Bruce: *Nessletter,* no. 54
Oct. 16, Mr. & Mrs. Crosbie: Searle, *Quarterly Newsletter,* Dec. 1982
Oct. 31, John Armstrong: ibid.
Nov. 8: ibid.
Nov. 20, Mr. & Mrs. Lees: ibid.
Dec. 9, J. Davidson: on exhibit in Searle's caravan

1983

Jan. 28, Roper: Searle, *Quarterly Newsletter,* Mar. 1983
Feb. 15, Ron Hyams: ibid.
*Feb. 23, P. Gilliat: ibid.
Mar. 24, F. Searle: ibid., June 1983
Apr. 4, Herb Macdonald: *Nessletter,* no. 57
Apr. 18, Mr. & Mrs. Lawton: Searle, *Quarterly Newsletter,* June 1983
May 15: ibid.
May 27, John Stansford: ibid.
June 18, Roy Jamieson: ibid., Sept. 1983
June 20, J. Nairn: *Nessletter,* no. 58
July 7, Mr. & Mrs. Williamson: Searle, *Quarterly Newsletter,* Sept. 1983
July 16: *Nessletter,* no. 60
July 25: Searle, *Quarterly Newsletter,* Sept. 1983
Aug. 6, James Newton: ibid., no. 60
*Aug. 6, Erik Beckjord: ibid., nos. 59–60; Searle, *Quarterly Newsletter,* Sept. 1983
Aug. 10, Mr. & Mrs. Downes: Searle, *Quarterly Newsletter,* Sept. 1983
Aug. 22, Mr. & Mrs. Marshall: ibid.
Aug. 24, Mrs. Jean Skelton: *Nessletter,* no. 60
*mid-Sept.: ibid., no. 64
Oct. 1, Hinley: Searle, *Quarterly Newsletter,* Dec. 1983
Nov. 17, Mike Noble: ibid.
Nov. 20, John Hall: ibid

1984

June 10, Jim Skelton: *Nessletter,* no. 64
Dec., Dr. G. Williamson: ibid., no. 68

1985

Mr. Walker: ibid.
Jan. 15, Jim Skelton: ibid.

Bibliography

To serve other students of the controversy over Loch Ness, I include here a more complete bibliography than is constituted by the references cited in the foregoing chapters. This bibliography is inevitably a compromise. Completeness is not to be attained on such a subject, for much material appears in newspapers (few of which are indexed) and in ephemeral bulletins and pamphlets that are rarely collected by libraries. Some compromise is necessary over the range of coverage of subject matter, which I have chosen to limit to items directly associated with Loch Ness and Loch Morar. If they exist, Nessies and Morags must be closely related to (at least one species of) sea serpents and probably also to some other lake monsters (if there are indeed others); but the literature is vast, one has to draw the line somewhere, and so I have excluded Ogopogo of Okanagan, Champ of Champlain, and so on, as well as sea serpents. A bibliography of material dealing more widely with freshwater monsters and with sea monsters has been published by George Eberhart; sea serpents are the subject of a definitive and well-documented book by Bernard Heuvelmans.

I have sought to make this bibliography, within its chosen scope, useful for more than a brief period of time, and so have focused on categories within which I could reasonably claim reasonably complete coverage; for example, I have listed items in newspapers only under those years for which my coverage of the particular papers has been comprehensive: the *Times* (London), the *New York Times,* and the *Glasgow Herald* (up to 1956), because indexes are available; the *Inverness Courier* from 1930 to 1939, thanks to the work of J. W. (Dick) MacKintosh of Inverness, who made an index for that paper from 1920 to 1939.

No one manner of organizing the bibliography would meet all likely requirements, so here again I have compromised. Books and sections in books are listed first and arranged by author, but for items in periodicals it seemed more generally useful to list by periodical, in view of the variety of authors who have written about the subject.

BOOKS

Akins, William. *The Loch Ness Monster.* New York: Signet (New American Library), 1977.

Baumann, E. D. *The Loch Ness Monster.* London and New York: Franklin Watts, 1972.

Bendick, Jeanne. *The Mystery of the Loch Ness Monster.* New York: McGraw-Hill, 1976 (for young children).

Berton, Jean. *Les monstres du Loch Ness et d'ailleurs.* Paris: France-Empire, 1977.

Binns, Ronald. *The Loch Ness Mystery Solved.* Shepton Mallet (Somerset): Open Books, 1983.

Burton, Maurice. *The Elusive Monster.* London: Rupert Hart-Davis, 1961.

Campbell, Elizabeth Montgomery (with David Solomon). *The Search for Morag.* London: Tom Stacey, 1972; New York: Walker, 1973.

Campbell, Steuart. *The Evidence about the Loch Ness Monster.* Wellingborough: Aquarian Press, 1986.

Cooke, David C., and Yvonne Cooke. *The Great Monster Hunt.* New York: W. W. Norton, 1969 (for children, grades 4–8).

Cornell, James. *The Monster of Loch Ness.* New York: Scholastic Book Services, 1977.

Costello, Peter. *In Search of Lake Monsters.* London: Garnstone, 1974; New York: Coward, McCann and Geoghegan, 1974; French ed., Paris: Plon, 1977.

Dinsdale, Tim. *The Leviathans.* London: Routledge and Kegan Paul, 1966; 2d rev. ed., London: Futura, 1976.

———. *Loch Ness Monster.* London: Routledge and Kegan Paul, 1961; Philadelphia: Chilton, 1962; 2d ed., Routledge and Kegan Paul, 1972; 3d ed., 1976; 4th ed., 1982.

———. *Monster Hunt* (1st rev. ed. of *The Leviathans*). Washington, D.C.: Acropolis, 1972.

———. *Project Water Horse.* London and Boston: Routledge and Kegan Paul, 1975.

———. *The Story of the Loch Ness Monster.* London: Allan Wingate and Target (Universal-Tandem), 1973 (for children).

Dunkling, Leslie. *The Mystery of the Loch Ness Monster.* Harlow (Essex): Longman, 1979 (for children—Longman Structural Readers Stage 1).

Gantès, Remy T. F. *Le mystère du Loch Ness.* Paris and Montreal: Etudes Vivantes, 1979 (for children).

Gould, Rupert T. *The Loch Ness Monster and Others*. London: Geoffrey Bles, 1934; New York: University Books, 1969.

Holiday, F. W. *The Dragon and the Disc*. Toronto: George J. McLeod, 1973; New York: W. W. Norton, 1973; London: Sidgwick and Jackson, 1973.

———. *The Great Orm of Loch Ness*. London: Faber and Faber, 1968; Toronto: George J. McLeod, 1969; New York: W. W. Norton, 1969; New York: Avon, 1970.

Mackal, Roy P. *The Monsters of Loch Ness*. Chicago: Swallow, 1976; London: Macdonald and Jane's, 1976.

Meredith, Dennis. *The Search at Loch Ness*. New York: Quadrangle (New York Times), 1977.

Perera, Victor. *The Loch Ness Monster Watchers*. Santa Barbara, Calif.: Capra Press, 1974.

Rabinovich, Ellen. *The Loch Ness Monster*. New York and London: Franklin Watts, 1979 (for young children).

Searle, Frank. *Nessie*. London: Coronet, 1976.

Smith, Warren. *Strange Secrets of the Loch Ness Monster*. New York: Zebra (Kensington), 1976.

Snyder, Gerald S. *Is There a Loch Ness Monster?* New York: Julian Messner, 1977 (for children); New York: Wanderer Books, 1977 (for younger children).

Thorne, Ian. *The Loch Ness Monster*. Mankato, Minn.: Crestwood House, 1978 (for children, grades 4–5).

Vibe, Palle. *Gaden I Loch Ness*. Copenhagen: Rhodos, 1970.

Whyte, Constance. *More than a Legend*. London: Hamish Hamilton, 1957; rev. 3d imp., 1961.

Witchell, Nicholas. *The Loch Ness Story*. Lavenham (Suffolk): Terence Dalton, 1974; 2d ed., 1976 (London: Book Club Associates, 1979); rev. ed., Harmondsworth (Middlesex): Penguin, 1975; rev. ed., London: Corgi (Transworld), 1982.

SECTIONS IN BOOKS

Akimushkin, Igor Ivanovich. *Sledy nevidannykh zverei* (The Tracks of Beasts Unseen). Moscow: Geografkis, 1961; German ed. *(Es gibt doch Fabeltiere)*, Leipzig: F. A. Brockhaus, 1963, pp. 223–38.

Alexander, Marc. *To Anger the Devil*. Suffolk (U.K.): Neville Spearman, 1978, pp. 13, 22–23, 67, 73–84, 110, 157, 166, 185–88.

Andrews, Roy Chapman. *All about Dinosaurs*. New York: Random House, 1953, pp. 128–29.

Barloy, Jean-Jacques. *Les animaux de la prehistoire*. Paris: France-Empire, 1978, pp. 199–206.

———. *Merveilles et mystères du monde animal*. Geneva: Famot, 1979, pp. 80–90.

———. *Serpent de mer et monstres aquatiques.* Geneva: Famot, 1978, pp. 53–73.

Bergier, Jacques, and Georges H. Gallet. *Le livre du mystère.* Paris: Albin Michel, 1975, pp. 147–54.

Bord, Janet, and Colin Bord. *Alien Animals.* London: Granada, 1980; Harrisburg, Pa.: Stackpole, 1981, pp. 1–43.

Boulenger, E. G. *A Natural History of the Seas.* New York: Appleton-Century, 1936, pp. 204–7.

Buehr, Walter. *Sea Monsters.* New York: W. W. Norton, 1966, chap. 5; New York: Archway (Pocket Books), 1971 (for children).

Burton, Maurice. *Animal Legends.* London: Frederick Muller, 1955, pp. 55–57, 108.

———. *Living Fossils.* London: Thames and Hudson, 1954, pp. 256, 262, 270.

———. *More Animal Legends.* London: Frederick Muller, 1959, pp. 171–87.

Carrington, Richard. *Mermaids and Mastodons.* New York: Rinehart, 1957, pp. 40–42.

Cazeau, Charles J., and Stuart D. Scott. *Exploring the Unknown.* New York: Plenum, 1979, pp. 217–22.

Cohen, Daniel. *The Encyclopedia of Monsters.* New York: Dodd, Mead, 1982, pp. 125–34.

———. *A Modern Look at Monsters.* New York: Dodd, Mead, 1970, pp. 91–126.

———. "Monsters." In George O. Abell and Barry Singer (eds.). *Science and the Paranormal.* New York: Charles Scribner's Sons, 1981, pp. 24–39.

———. *Myths of the Space Age.* New York: Dodd, Mead, 1965, pp. 5, 198–211.

Corliss, William R. *Incredible Life.* Glen Arm, Md.: Sourcebook Project, 1981, pp. 308–18.

———. *Mysteries Beneath the Sea.* New York: Thomas Y. Crowell, 1970, pp. 138–55.

———. *Strange Life.* Glen Arm, Md.: Sourcebook Project, vol. B-1, 1976, pp. 32, 244–55.

Darling, Lois, and Louis Darling. *The Sea Serpents Around Us.* Boston: Little, Brown, 1965, pp. 14–35 (a picture book for young children).

Ditmars, Raymond Lee. *Confessions of a Scientist.* Freeport, N.Y.: Books for Libraries Press, 1934 and 1970, pp. 205–17.

Dunkel, Ulrich. *Abenteuer mit Seeschlangen.* Stuttgart: Kreuz, 1961, pp. 99–125, 180–96.

Eberhart, George M. *Monsters—A Guide to Information on Unaccounted For Creatures, Including Bigfoot, Many Water Monsters, and Other Irregular Animals.* New York and London: Garland, 1983, pp. 33–80.

Farson, Daniel, and Angus Hall. *Mysterious Monsters*. London: Aldus Books, 1978; New York: Mayflower Books, 1979, pp. 182–99.

Fromentin, Pierre. *Monstres et bêtes inconnues*. Marne: Tours, 1948 and 1954, pp. 39–44.

Godwin, John. *This Baffling World*. New York: Hart, 1968, pp. 312–21, 328.

Gould, Rupert T. *The Stargazer Talks*. London: Geoffrey Bles, 1944, pp. 85–88 (reprinted as *More Oddities and Enigmas*. New York: University Books, 1973).

Graves, Robert, and Alan Hodge. *The Long Week-End*. New York: Macmillan, 1941, pp. 277–78; London: Faber and Faber, 1950, pp. 288–89, 310–11.

Guenette, Robert, and Frances Guenette. *The Mysterious Monsters*. Los Angeles: Sun Classic, 1975, pp. 17–32.

Haining, Peter. *Ancient Mysteries*. New York: Taplinger, 1977, pp. 137–57.

Hall, Angus. *Monsters and Mythic Beasts*. London: Aldus Books, 1975, pp. 62–83; Garden City, N.Y.: Doubleday, 1976; also Danbury Press, n.d., in a series titled *The Supernatural*.

Helm, Thomas. *Monsters of the Deep*. New York: Dodd, Mead, 1962, pp. 182–87.

Hennig, Richard. *Wo lag das Paradies? Raetselfragen der Kulturgeschichte und Geographie*. Berlin: Tempelhof, 1950, pp. 277–90, 305; French ed., *Les grandes enigmes de l'universe*. Robert Laffont, 1957, pp. 245–54.

Heuvelmans, Bernard. *In the Wake of the Sea-Serpents*. New York: Hill and Wang, 1968, passim.

———. *Les derniers dragons d'Afrique*. Paris: Plon, 1978, pp. 87–91, 93–102, 162, 186, 225, 259, 281, 311, 325, 352, 371.

Hitching, Francis. *The Mysterious World: An Atlas of the Unexplained*. New York: Holt, Rinehart and Winston, 1979, pp. 197–201, 247.

Keel, John A. *Strange Creatures from Time and Space*. Greenwich, Conn.: Fawcett, 1970, pp. 268–71.

Krumbiegel, Ingo. *Von neuen und unentdeckten Tierarten*. Stuttgart: Kosmos (Franckh'sche Verlagshandlung), 1950, pp. 55–56.

Landsburg, Alan. *In Search of Myths and Monsters*. New York: Bantam, 1977, pp. 48–89.

———. *In Search Of. . . .* New York: Everest House, 1978, pp. 215–23.

Lange, P. Werner. *Seeungeheuer: Fabeln und Fakten*. Leipzig: F. A. Brockhaus, 1979, pp. 94–101.

Laycock, George. *Strange Monsters and Great Searches*. Garden City, N.Y.: Doubleday, 1973, pp. 6–14; London and Sydney: Pan Books, 1976 (for children).

———. *Mysteries, Monsters and Untold Secrets*. Garden City, N.Y.: Doubleday, 1977, pp. 9–20.

Lee, John, and Barbara Moore. *Monsters among Us: Journey to the Unexplained.* New York: Pyramid, 1975, pp. 63–76.

Lethbridge, T. C. *Ghost and Divining-Rod.* London: Routledge and Kegan Paul, 1963, pp. 114–22.

Ley, Willy. *Exotic Zoology.* New York: Viking, 1959; New York: Capricorn, 1966, pp. 234–36.

———. *The Lungfish and the Unicorn.* New York: Modern Age Books, 1941, pp. 76, 99–102.

———. *The Lungfish, the Dodo and the Unicorn.* New York: Viking, 1948, pp. 114, 125–27.

MacDougall, Curtis D. *Hoaxes.* New York: Macmillan, 1940, pp. 13–14.

Mackal, Roy P. *Searching for Hidden Animals.* Garden City, N.Y.: Doubleday, 1980, passim.

Maine, C. E. "The Shy Serpent of Loch Ness" and "Denizens of the Deep Lakes." In John Canning (ed.). *Fifty True Mysteries of the Sea.* New York: Stein and Day, 1980, pp. 197–219.

McEwan, Graham J. *Sea Serpents, Sailors and Sceptics.* London and Boston: Routledge and Kegan Paul, 1978, pp. 106–9.

McHargue, Georgess. *The Beasts of Never.* Indianapolis: Bobbs-Merrill, 1968, pp. 99–107.

Michell, John, and Robert J. M. Rickard. *Living Wonders.* London: Thames and Hudson, 1982, pp. 20–21, 23–25.

———. *Phenomena: A Book of Wonders.* New York: Pantheon, 1977, pp. 118–19.

Moon, Mary. *Ogopogo.* Vancouver: J. J. Douglas, 1977, pp. 8–10.

Newton, Michael. *Monsters, Mysteries and Man.* Reading, Mass.: Addison-Wesley, 1979, pp. 91–102 (for older children).

Peachment, Brian. *You Be the Judge.* London: Edward Arnold, 1972, pp. 105–35.

Poulsen, C. M. *Soeslangens Gade* (Enigma of the Sea-Serpent). Copenhagen: H. Hirschsprung, 1959, pp. 49–100.

Quinn, Daniel. *Land and Sea Monsters.* Northbrook, Ill.: Hubbard Press, 1971, pp. 40–47.

Reader's Digest. *Into the Unknown.* Pleasantville, N.Y.: Reader's Digest Association, 1981, pp. 112–13.

———. *Mysteries of the Unexplained.* Pleasantville, N.Y., and Montreal: Reader's Digest Association, 1982, pp. 141–51.

———. *Strange Stories, Amazing Facts.* Pleasantville, N.Y., and Montreal: Reader's Digest Association, 1976, pp. 424–27.

Resciniti, Angelo, and Duane Damon. *Big Foot and Nessie—Two Mysterious Monsters.* Columbus, Ohio: School Book Fairs, 1979, pp. 68–125.

Rickard, Robert, and Richard Kelly. *Photographs of the Unknown.* London: Book Club Associates and Granta, 1980, pp. 14–21.

Russell, Eric Frank. *Great World Mysteries.* London: Dennis Dobson, 1957, pp. 147–49.

Sanderson, Ivan T. *Investigating the Unexplained.* Englewood Cliffs, N.J.: Prentice-Hall, 1972, pp. 5–38.

Shine, Adrian. "Loch Ness Monster." In Peter Brookesmith (ed.). *Creatures from Elsewhere.* London: Orbis, 1982, pp. 58–71 (reprinted from *The Unexplained,* nos. 11–14, 1980–81).

Sladek, John. *The New Apocrypha.* London: Hart-Davis, MacGibbon, 1973, pp. 84–88.

Spencer, John Wallace. *Limbo of the Lost.* New York: Bantam, 1973, pp. 97–100; rev. ed., *Limbo of the Lost Today,* Springfield, Mass.: Phillips, 1975 and 1979, pp. 194–98.

Spraggett, Allen. *New Worlds of the Unexplained.* New York: Signet, 1976, pp. 11–12.

Sweeney, James B. *A Pictorial History of Sea Monsters and Other Dangerous Marine Life.* New York: Nelson-Crown, 1972, pp. 118–22; New York: Bonanza, 1976.

———. *Sea Monsters—A Collection of Eyewitness Accounts.* New York: David McKay, 1977, pp. 80–88.

Thurber, James. *Alarms and Diversions.* New York: Harper and Bros., 1957, pp. 52–65.

Welfare, Simon, and John Fairley. *Arthur C. Clarke's Mysterious World.* New York: A & W Publishers, 1980, pp. 100–115.

Wignall, Sydney. *In Search of Spanish Treasure.* Newton Abbot (Devon, U.K.) and North Pomfret, Vt.: David and Charles, 1982, pp. 232–36, 241–42.

———. "Quest for a Surviving Dinosaur." *Spirit of Enterprise* (1984 Rolex Awards). London: Aurum Press for Montres Rolex S.A., 1984, pp. 268–69.

Wilson, Colin. *Enigmas and Mysteries.* Garden City, N.Y.: Doubleday, 1976, pp. 78–82.

World Almanac. *The World Almanac Book of the Strange.* New York: Signet (New American Library), 1977, pp. 292–99.

Zug, George. "Once More into the Loch." *Encyclopedia Britannica Yearbook of Science and the Future,* 1978, pp. 154–69.

ARTICLES

Aesculape (Paris), Nov. 1961, 44:35–37; Jean Boullet, "Le monstre du Loch Ness."

Amateur Photographer (London), 21 Jan. 1976; Iain A. Meecham, "Glint in the Eye for Nessie" (letter).

Animals, 30 July 1962; Maurice Burton, "Is This the Loch Ness Monster?"

———, 24 Mar. 1964, 3 : 448; Maurice Burton, "The Loch Ness Monster."
———, Aug. 1966; review of Dinsdale, *The Leviathans* (1966).
———, Sept. 1966, 9 : 259; comment.
Anomaly (publ. John Keel, New York), no. 8, Summer 1972; Henry Allen, an interview with Tim Dinsdale.
Antiquity, 8 (1934): 85–86; by C., O.G.S.
Argosy, Dec. 1970, 46–49; Ivan T. Sanderson, "Two Loch Ness Monsters?"
———, Nov. 1976, 39–42; Jack Carroll, "The One Man Who Tracked and Sighted (Nessie) for Seven Years."
Argosy Special: Monsters, 1977, 27–31; Jack Carroll, "Loch Ness Monster" (reprint from *Argosy*, Nov. 1976).
Atlantic, June 1962, 116; review of Dinsdale, *Loch Ness Monster* (1961).
———, 222, no. 4 (1968): 150; Phoebe Adams, review of Heuvelmans, *In the Wake of the Sea-Serpents* (1968).
———, Jan. 1970, 42–47; John McPhee, "Pieces of the Frame."
———, Sept. 1974; Phoebe Adams, review of Costello, *In Search of Lake Monsters* (1974).
Atlantic Advocate, Apr. 1969; Bruce S. Wright, "Watch for Monsters" (review of Heuvelmans, *In the Wake of the Sea-Serpents* [1968]).
Audubon, 71 (1969): 90–91; Roger Caras, "What Unknown Creatures of the Depths?" (review of Heuvelmans, *In the Wake of the Sea-Serpents* [1968]).
Australian Museum Magazine, 16 Apr. 1934, 204–8.
Best Sellers, Aug. 1976, 36 : 174; K. R. Mullin, review of Mackal, *The Monsters of Loch Ness* (1976).
Bioscience, Dec. 1977, 817; J. Richard Greenwell, "Seeking Nessie" (review of Meredith, *The Search at Loch Ness* [1977]).
Booklist, 1 Jan. 1969, 65 : 469; review of Heuvelmans, *In the Wake of the Sea-Serpents* (1968).
Boston Magazine, Feb. 1977, 69 : 64–65,82–92; Peter Birge, "The Enhanced Vision of Robert H. Rines."
Boys' Life, Oct. 1977, 27–29; William Wise, "Nessie—The Mysterious Loch Ness Monster."
———, Nov. 1980, 28, 30, 66–67; George Laycock, "The Monster of Loch Ness."
———, July 1984, 24–27; David Ballard, "A Bonny Day on the Loch."
British Journal of Photography, 20 Apr. 1984, 402–5, 410; Steuart Campbell, "The Surgeon's Monster Hoax."
———, 18 May 1984, 505; Ian Johnson, "Nessie Otter No?"
CalTech News, Dec. 1975; "Computer Aids in Search for 'Nessie'."
Catholic Digest (St. Paul, Minn.), 37, no. 10 (Aug. 1973): 66–71; Ray Bronikowski, "Loch Ness Monastery" (condensed from "Our Sunday Visitor").

Catholic Weekly (Australia), 11 Dec. 1975, 24–25, 30; Alan Gill, "The Loch Ness Monster: Fish, Flesh or Good Red Herring?"

Changing Times, 34, no. 8 (Aug. 1980): 65; review of Mackal, *The Monsters of Loch Ness* (1976).

Chemical and Engineering News, 31 May 1971, 25; "Loch Ness Monster Pursued with Scents."

———, 31 Jan. 1977, 40; "Loch Ness Monster—No Drop in Interest."

———, 24 Sept. 1979, 66; "Scientists Say Loch Ness Monster May Be Elephants."

Chemistry, Nov. 1969, 5; "On the Difficult Art of Monster Stalking."

Choice, Mar. 1969, 6:83; review of Heuvelmans, *In the Wake of the Sea-Serpents* (1968).

———, Dec. 1974, 11 : 1492; review of Costello, *In Search of Lake Monsters* (1974).

———, Sept. 1976, 13 : 847; review of Mackal, *The Monsters of Loch Ness* (1976).

———, Nov. 1976, 13 : 1158; review of Dinsdale, *The Leviathans* (1976).

Circus Magazine, 31 Mar. 1977, 54–57; Lynn Hudson, "The Man Who Saw the Monsters."

Collier's, 30 June 1934, 22, 34; Corey Ford, "Sea-Serpent Control."

Columbia Journalism Review, Sept./Oct. 1976, 42–44; Jon Swan, "Monster Swamps 'Times'."

Commonweal, 16 Feb. 1934, 424; "Monstrum Informe, etc."

———, 20 Apr., 677–78; David Hunter Blair, "Elusive Monster of Loch Ness."

Contemporary Review, Sept. 1974, 225 : 138–44; Richard Whittington-Egan, "Loch Ness: The Monstrous Zoological Problem."

———, Oct. 1975, 227:221–22; Richard Whittington-Egan, "The Monster Under Scrutiny" (reviews of Dinsdale, *Project Water Horse*, [1975] and Witchell, *The Loch Ness Story* [1974, 1975]).

Cryptozoology, 1 (1982): 40–45; Henry H. Bauer, "The Loch Ness Monster: Public Perception and the Evidence."

———, 2 (1983): 98–102; James E. King and J. Richard Greenwell, "Attitudes of Biological Limnologists and Oceanographers Toward Supposed Unknown Animals in Loch Ness."

Discover, Oct. 1982, 15; "Pining Over Nessie."

———, Sept. 1984, 6; "Skeptical Eye—The (Retouched) Loch Ness Monster."

———, Mar. 1985, 35–42; Jared Diamond, "In Quest of the Wild and Weird."

Discovery, Jan. 1934, 15 : 14; "The Loch Ness 'Monster'."

Diver (U.K.; formerly *Triton*), Sept.–Oct. 1969, 144–45; Arthur Bourne, "In the Monster's Lair."

Doubt, no. 16 (1946): 237; "No Such Animal."

———, no. 20 (1948): 303; "Run of the Mill."

———, no. 30 (1950): 36–37; "No Such Animal."
———, no. 38 (1952): 171; "Nessy's Pa Mort."
Economist, 19 Apr. 1969, 73–74; "Sea Beasts Known and Unknown" (review of Heuvelmans, *In the Wake of the Sea-Serpents* [1968]).
———, 28 Nov. 1970; "Monsters—How Do You Know They Aren't Real?"
———, 15 July 1972, 55; "Giant Swimmers" (review of Campbell and Solomon, *The Search for Morag* [1972], and Dinsdale, *Loch Ness Monster* [1972]).
———, 13 Dec. 1975, 33; "Loch Ness: A Monstrous Doubt."
———, 7 May 1983, 66 (U.K. ed., p. 46); "The Nessiesary Monster."
EDN, 5 Nov. 1976, 21–23; Paul Snigier, "Three-channel Side-scan Sonar Detects Loch Ness 'Monsters' and Ancient Ruins."
EGG Ink, issue 3 (1974): 18–19; Walter A. Deane, "The Loch Ness Monster Is Real."
Elks Magazine, June 1966, 56; "The Monster of Loch Ness" (editorial).
Equinox, Sept. 1982; Virginia Morell, "He Hunts for Living Dinosaurs."
European Community, Apr./May 1976, 34–40; Walter Sturdivant, "Loch Ness Monster."
European Scientific Notes (Office of Naval Research, London), ESN–23–2 (28 Feb. 1969): 36–38; D. C. Hornig, "Fabulous Monster Confirmed by Sonar?"
Everybody's, 21 Feb. 1959; Alex M. Campbell (as told to Graham Fisher), "The Day I Saw the Loch Ness Monster."
Everybody's Weekly, 25 Apr. 1959; George Langelaan, "Is the Loch Ness Monster French?"
L'Express (France), 7 June 1976, 98–101; "La science s'attaque au Loch Ness."
Fate, 7, no.12 (1954): 6; "I See by the Papers—Time of the Monsters."
———, 8, no. 5 (1955): 10–11; "Loch Ness Monster Again."
———, 9, no. 4 (1956): 71; "The Loch Ness Monster Returns."
———, 11, no. 6 (1958): 61–64; Constance Whyte, "Yes, There Is a Loch Ness Monster."
———, 11, no. 9 (1958): 124–25; Maude Kapierlian, "Loch Ness Serpents."
———, 12, no. 7 (1959): 6, 8; "I See by the Papers—Monster, Monster, Everywhere."
———, 13, no. 10 (1960): 16, 18; "About Some Little Beasties."
———, 14, no. 5 (1961): 82–84; Paul Foght, "The Shy Monster of Loch Ness."
———, 15, no. 9 (1962): 10; "Ach Loch Ness."
———, 16, no. 7 (1963): 23–4; "I See by the Papers—Yes, There Is a Monster."
———, 16, no. 10 (1963): 77; "News from Loch Ness."
———, 17, no. 11 (1964): 99–110; Ivan T. Sanderson, "The Big Lake Monster Hunt."

———, 18, no. 2 (1965): 26; "I See by the Papers—Note on Loch Ness."
———, 18, no. 11 (1965): 24, 26; "Och! Loch Ness!"
———, 19, no. 8 (1966): 24; "They Mean Business."
———, 19, no. 9 (1966): 28; "Last Word from Loch Ness."
———, 19, no. 12 (1966): 34–43; Ivan T. Sanderson, "R.A.F. Report on Loch Ness Monster Photos."
———, 20, no. 2 (1967): 18, 20; "Missed Again."
———, 20, no. 11 (1967): 22, 24–25; "I See by the Papers—The Never-Ending Hunt."
———, 21, no. 1 (1968): 22–24; "I See by the Papers—Hunt at Loch Ness; A Veritable Herd."
———, 21, no. 5 (1968): 26, 28, 30; "Meanwhile Back at the Loch."
———, 21, no. 12 (1968): 52–54; Betty Lou White, "Rethinking the Loch Monster."
———, 22, no. 1 (1969): 26, 28, 30, 32; "The Loch Ness Summer."
———, 22, no. 4 (1969): 10; "Sonar at Loch Ness."
———, 23, no. 5 (1970): 12, 14; "Hail to Nessie."
———, 23, no. 11 (1970): 28, 30; "Monster Report: Loch Ness."
———, 24, no. 4 (1971): 32, 34, 35; Curtis Fuller, "I See by the Papers—Round Loch Ness; Meanwhile On to Loch Morar."
———, 25, no. 10 (1972): 39; Curtis Fuller, "I See by the Papers—Loch Ness Hoax."
———, 26, no. 6 (1973): 35–36; "A Photo of Nessie?"
———, 27, no. 4 (1974): 16, 18, 19; "'Kussie' vs. 'Nessie'."
———, 29, no. 8 (1976): 26, 28; Curtis Fuller, "I See by the Papers—Nessie Still an Enigma."
———, 29, no. 9 (1976): 14, 16, 18; "What's New on Nessie?"
———, 30, nos. 9 and 10 (1977): 36–43, 68–74; Jerome Clark, "Tracking the Loch Ness Monsters."
———, 31, no. 5 (1978): 117–19; Murray Carr, "Peals of Laughter."
———, 32, no. 9 (1979): 25; "I See by the Papers—Dolphins' Search."
———, 33, no. 3 (1980): 87; Jerome Clark, "Update."
———, 33, no. 4 (1980): 23–24, 26; "I See by the Papers—What's with Monsters?"
———, 34, no. 2 (1981): 85; Jerome Clark, "Update—It Can't Be; Therefore It Isn't—Move Over, Nessie."
———, 35, no. 11 (1982): 26, 28; "A Gaggle of Monsters."
———, 37, no. 2 (1984): 30–32; "While at Loch Ness. . . ."
———, 38, no. 1 (1985): 103–4; Gordon Stein, review of Binns, *The Loch Ness Mystery Solved* (1983).
Field, 4 Nov. 1933, 162:1147; J. E. Hamilton, "The Loch Ness Monster and Sea Serpents in General."
———, 25 Nov. 1933, 162:1371; 16-Bore, "Loch Ness Monster" (letter).
———, 2 Dec. 1933, 162:1431; George L. Harrison, "The Loch Ness Monster" (letter).

———, 13 Jan. 1934, 163:77–78; (letters) C.B.E., "Monsters and Skeptics"; Kelpie, "Another Monster Suggestion."
———, 27 Jan. 1934, 163:167; Martin A. C. Hinton, "Summing-up the Loch Ness Monster."
———, 10 Feb. 1934, 163:276; "Evidence from Loch Ness."
———, 10 Feb. 1934, 163:289; (letters) E.C., "Against the Seal Theory"; Courtenay Bennett, "Monster or Eel?"; David Hunter Blair, "The Loch Ness 'Monster'"; Herbert Maxwell, "Highland Superstition"; Portland, "The Loch Ness 'Monster'"; J. P. Kay Robinson, "African Apparitions."
———, 28 Apr. 1934, 163:936; "Loch Ness."
———, 26 May 1934, 163:1245; R. A. R. Meiklem, "The Loch Ness Problem" (letter).
———, 2 June 1934, 163:1305; B. A. Russell, "The Loch Ness Problem" (letter).
———, 22 Sept. 1934, 164:668–69; Edward M. Mountain, "Solving the Mystery of the Loch Ness Monster."
———, 13 Oct. 1934, 164:871; (letters) W. T. Calman, D. Seth-Smith, A. Ezra, B. Barnett, F. C. Fraser, "The Loch Ness Film."
———, 20 Oct. 1934, 164:928; M. A. C. Hinton, "The Loch Ness Film."
———, 27 Oct. 1934, 164:979; (letters) C. E. Radclyffe and "Salar," "The Loch Ness Monster."
———, 3 Nov. 1934, 164:1029; Gilfrid Hartley, "The Loch Ness Monster" (letter).
———, 10 Nov. 1934, 164:1089, C. M. Hope, "Loch Ness and Its Monster" (letter).
———, 16 Feb. 1935; C. E. Radclyffe, "The Loch Ness Monster" (letter).
———, 28 July 1951; J. Harper Smith, "The Loch Ness Monster" (letter).
———, 5 Mar. 1959; F. W. Holiday, "Truth and the Monster."
———, 23 Nov. 1961, 951–53; David James, "Time to Meet the Monster."
———, 14 June 1962, 1160; David James, "The Monster Again."
———, 1 Nov. 1962; F. W. Holiday, "The Monster: Another View."
———, 8 Dec. 1966, 1216; F. W. Holiday, "Only a Worm in Loch Ness?"
———, 21 Aug. 1969, 369; H. H-B., "A Monster-Hunt Warms Up."
———, 16 Apr. 1970, 683; "A Creature in Loch Ness?"
———, 26 Nov. 1970, 1113–14; David James, "Closing in on the Loch Ness Monster."
———, 29 July 1971; "Great Orm" (review of Holiday, *The Great Orm of Loch Ness* [1968]).
———, 10 Aug. 1972; Christopher Carter, "Nessie's Rival" (review of Campbell, *The Search for Morag* [1972], and Dinsdale, *Monster Hunt* [1972]).

———, 23 Oct. 1975, 769–70; David James, "Life in the Deep in Loch Ness."
———, 27 Nov. 1975; Alan Wilkins, "Monster: The Four Vital Sightings."
———, 4 Dec. 1975; Alan Wilkins, "The Shapes on the Loch."
———, 5 Feb. 1976, 204–5; F. W. Holiday, "The Case for a Spineless Monster."
Filmmakers Newsletter, June 1976, 30–35; Charles W. Wyckoff, "Filming the Loch Ness Monster."
Flying Saucer Review, Sept.–Oct. 1961, 7:15; "The Loch Ness Monster: An Old Friend."
———, Sept.–Oct. 1971, 17:12–14; F. W. Holiday, "Monsters and UFOs: Some Observations on Loch Ness."
———, Jan.–Feb. 1972, 18:29; Peter Jackson, "Monster and UFOs" (letter).
———, Sept.–Oct. 1972, 18:22–24; F. W. Holiday, "Science and Loch Ness."
———, Jan.–Feb. 1973, 19:15–17; F. W. Holiday, "Dragons and UFOs at Loch Ness."
———, Sept.–Oct. 1973, 19:3–7, 13; F. W. Holiday, "Exorcism and UFO Landing at Loch Ness."
———, Apr. 1975, 20:7–10; F. W. Holiday, "A Brief Taste of Fairyland."
———, Aug. 1975, 21:26–27; John M. Lade, "An Informed Speculation."
———, Feb. 1976, 30–31; F. W. Holiday, "Facts for 'Informed Speculators'."
The Fortean (Fortean Society Magazine), no. 10 (Autumn 1944): 145; "No Such Animal."
Fortean Society Magazine, no. 1 (Sept. 1937): 4; "Loch Ness Litters."
Fortean Times, no. 14 (Jan. 1976): 11–16; R. J. M. Rickard, "The Emperor's New Monster."
———, no. 15 (Apr. 1976): 12; "Loch Ness, Inverness."
———, no. 19 (Dec. 1976): 12–17; "Unidentifieds."
———, no. 21 (Spring 1977): 27–30; "Unidentifieds."
———, no. 22 (Summer 1977): 18–26; "Nessie: Unidentifieds."
———, no. 23 (Autumn 1977): 20–22; Doc Shiels, "Words from the Wizard."
———, no. 24 (Winter 1978): 14–16; "Unidentifieds—GSW Examines Doc's Nessie Photos."
———, no. 25 (Spring 1978): 49–50; Thomas E. Bearden, "Nessie Photos and the New Physics" (letter).
———, no. 29 (Summer 1979): 26–31, back cover; "Nessie: The Shiels 1977 Photos."
———, no. 30 (Autumn 1979): 28–30; "Nessie Notes."

———, no. 30 (Autumn 1979): 67; Mike Crowley, "Nessie the Elephant?" (letter).
———, no. 33 (Autumn 1980): 36–37; "Unidentifieds—Loch Ness 3 Sept. 1978."
———, no. 42 (1984): 62–67; Doc Shiels, "Mother Nature's Jumbo Jet."
Geo, May 1981, 3 : 148; "Water Monsters Are Shown to Be Mirages."
The Grocer (U.K.), 29 Sept. 1984; "Monster Man Takes to the Water."
Guardian Weekly, 20 Mar. 1969, 14; C. M. Yonge, "Great Unknowns" (review of Heuvelmans, *In the Wake of the Sea-Serpents* [1968]).
Hansard (House of Commons Proceedings), 12 Dec. 1933, 177–79.
———, 20 May 1958, 1093–94.
———, 304, no. 95 (16 July 1969): 262–64; "Loch Ness Monster: Submarine Research."
Harper's, Feb. 1956, 24–25; "Personal and Otherwise."
———, Feb. 1956, 35–37; David M. Slorach, "How I Met the Loch Ness Monster."
———, June 1977, 84; review of Meredith, *The Search at Loch Ness* (1977).
Herpetological Review, June 1976, 41–46; Kraig Adler, "Loch Ness Monster Evidence Presented at Cornell University."
Holiday, Sept. 1957, 54–55, 118, 120–22, 124, 126; James Thurber, "There's Something Out There!"
Horoscope, Sept. 1977, 43 : 22–23, 26, 28; George Whitelaw, "The Mystery of the Loch Ness Monster Revealed!"
Humanist (London), 76/77 (1961): 362–67: Pat Sloan, "The Loch Ness Monster."
———, July/Aug. 1977, 44–45; Philip J. Klass, "The Mermaid Investigation Sightings Society (MISS)."
———, Sept./Oct. 1980, 61–62; Kenneth S. Saladin, "The Loch Ness Monster: A Mirage?"
IEEE Spectrum, 15, no. 2 (Feb. 1978): 26–29; Harold E. Edgerton and Charles W. Wyckoff, "Loch Ness Revisited."
Illustrated London News, 11 Nov. 1933, 760–61; W. P. Pycraft, "Loch Ness in Possession of a 'Sea-Serpent'!"
———, 6 Jan. 1934, 4; G. K. Chesterton, "Our Notebook."
———, 6 Jan. 1934, 8–9; "The Loch Ness Monster Paralleled in Canada: 'Cadborosaurus'."
———, 13 Jan. 1934, 39–42.
———, 28 Apr. 1934, 650.
———, 5 May 1934, 701.
———, 11 Aug. 1934, 219; C.K.A., "Leviathan" (review of Gould, *The Loch Ness Monster and Others* [1934]).
———, 18 Aug. 1934, 261; "The Loch Ness 'Monster' Seen Twenty-one Times in Four Weeks."
———, 1 Sept. 1934, 315; "'Monsters' Caught and Not Yet Caught: A Record Tunny and—What?"

———, 15 Dec. 1934, 1011; "A Canadian 'Monster': Sea-Cow, Basking Shark, or 'Cadborosauras'?"
———, 8 Dec. 1951, 950; Maurice Burton, "The Mystery of Loch Ness."
———, 8 Aug. 1953, 222; Maurice Burton, "Eel Watching."
———, 8 Jan. 1955, 66–67.
———, 15 June 1957, 992; Maurice Burton, "Monsters: In Fact and (?) Fiction."
———, 20 Feb. 1960, 316; Maurice Burton, "The Loch Ness Monster."
———, 16, 23, 30 July 1960, 110–11, 150–52, 192–93; Maurice Burton, "The Problem of the Loch Ness Monster: A Scientific Investigation."
———, 26 Nov. 1960, 956; Maurice Burton, "The Tantalizing Monster."
———, 27 May 1961, 896; Maurice Burton, "Loch Ness Monster: A Burst Bubble?"
———, 24 Feb. 1962, 300; Maurice Burton, "Loch Ness: An End to the Plesiosaur Story."
———, Xmas 1978, 31–35; Richard Gordon, "The Immortal Nessie."
L'Illustration (Paris), 20 Jan. 1934; "Les surprenants aspects du monstre du Loch Ness, d'après les temoins qui l'auraient aperçu."
In Britain, Apr. 1973, 14–18; "The Shyest Prima Donna in the World"; and James Lauder, "The Minor Monsters."
L'Inexpliqué (Paris; abridged version of *The Unexplained* [London]), no. 9 (1981): 166–69; "La mysterieuse nageoire du Loch Ness."
INFO Journal, no. 5 (Fall 1969): 24–26; "Loch Nessiana."
———, no. 6 (Spring 1970): 27; Tim Dinsdale, "Report from Loch Ness."
———, no. 11 (Summer 1973): 18–20; "Loch Ness Monster Evidence Is Mounting—Holiday Writes."
———, no. 14 (Nov. 1974): 9–11; F. W. Holiday, "About Monsters and Such."
———, no. 16 (Mar. 1976): 2–4; F. W. Holiday, "Water Monsters: The Land Sighting Paradox."
———, no. 20 (Nov. 1976): 2–4; F. W. Holiday, "The Great Serpent Mystery."
———, no. 20 (Nov. 1976): 14; "Lake Monsters."
———, no. 21 (Jan. 1977): 14; "Newsclips—Nessie."
———, no. 30 (July–Aug. 1978): 10; "'Nessie' Makes an Appearance."
International Wildlife, July–Aug. 1971, 17–19; Janice and Charles Robbins, "Is There Really a Loch Ness Monster?"
———, 6, no. 2 (1976): 38–39; Jim Doherty, "The Real Nessie?"
JAMA (previously *Journal of the American Medical Association*), 14 Sept. 1970, 213 : 1884–85; G. G. Liddle, "The Loch Ness Monster: Report of a Case Without Review of the Literature."
Journal des Débats, 26 Jan. 1934, 41 : 152; V., "Variétés Scientifiques—La fin du monstre du Loch Ness."

———, 12 Jan. 1934, 41:69–71; Henry de Varigny, "La vie scientifique—Revue de sciences—Le monstre du Loch Ness."
Journal of Irreproducible Results, July 1974, 20:15–18.
King's College Hospital Gazette, no. 38, Spring 1950; vol. 29, no. 1, 6–18; "The Loch Ness Monster" (no identification of author is given, but attributed to Constance Whyte by Witchell, *The Loch Ness Story* [1974]).
Kodak News, 4, no. 7 (2 Apr. 1971): 5; "That Loch Ness Mystery: Photography Will Help to Solve It."
Ladies Home Journal, Apr. 1971; Mary Fiore, "How I Learned to Love the Loch Ness Monster."
De levende Natuur (Amsterdam), 12 Nov. 1934–1 Feb. 1935, 39: 214–25, 248–255, 281–87, 316–20; A. C. Oudemans, "Het Loch Ness Monster."
Library Journal, July 1962, 2563; Aaron L. Fessler, review of Dinsdale, *Loch Ness Monster* (1961).
———, Aug. 1968, 2889–90; Daniel M. Simms, review of Heuvelmans, *In the Wake of the Sea-Serpents* (1968).
———, 1 May 1969, 1890; R. E. Swinburne, Jr., review of Holiday, *The Great Orm of Loch Ness* (1968).
———, 15 Mar. 1970; A. C. Haman, review of Cooke and Cooke, *The Great Monster Hunt* (1969).
———, 1 May 1973, 1500; Jonathan F. Husband, review of Campbell and Solomon, *The Search for Morag* (1972).
———, 1 Nov. 1973; Robert Molyneux, review of Holiday, *The Dragon and the Disc* (1973).
———, Aug. 1974; Jonathan F. Husband, review of Costello, *In Search of Lake Monsters* (1974).
———, 1 Apr. 1976; Jonathan F. Husband, review of Mackal, *The Monsters of Loch Ness* (1976).
———, 1 Apr. 1977; Jonathan F. Husband, review of Meredith, *The Search at Loch Ness* (1977).
Life, 20 Sept. 1963, 17–18; Paul Mandel, "Showdown for Loch Ness."
Limnology and Oceanography, 17 (1972): 796–97; R. W. Sheldon and S. R. Kerr, "The Population Density of Monsters in Loch Ness."
———, 18 (1973): 343–46; W. Scheider and P. Wallis, "An Alternate Method of Estimating the Population Density of Monsters in Loch Ness"; C. H. Mortimer, "The Loch Ness Monster—Limnology or Paralimnology?"; R. W. Sheldon and S. R. Kerr, "Reply to Comments of C. H. Mortimer."
Listener, 8 Nov. 1933, 690–91; Philip A. Stalker, "The Monster of Loch Ness."
———, 10 Mar. 1937, 463–64; E. G. Boulenger, "Aquatic Monsters—Do They Exist?"
———, 7 Apr. 1937, 657–59; R. T. Gould, "Sea Monsters—A Vindication."

———, 1 Aug. 1968, 143; "Out of the Air—Monstrous Behaviour."
Literary Digest, 27 Jan. 1934, 18; "The Crop of Sea-Serpents."
MCZ Newsletter (Museum of Comparative Zoology, Harvard), 5, no. 2 (Winter 1976); "Is There or Isn't There an Aquatic Animal Residing in Loch Ness?"
Machine Design, 14 Sept. 1967, 44–52; Leo F. Spector, "The Great Monster Hunt."
———, 41, no. 16 (10 July 1969): 42; Leo Spector, "The Little Yellow Monster-Chasing Submarine."
MacLean's, 12 Aug. 1961, 1; Dorothy Eber, "The Scientific Search for a Prehistoric Monster."
———, 6 Sept. 1976, 89 : 38–46; Michael Enright, "Waiting for Nessie."
Mayfair, 8, no. 6 (1973): 16–18, 37, 38, 87; Nick Witchell, "The New Side to the Loch Ness Monster: Five Times It's Been Seen on Land."
Midwest Magazine (Sunday newspaper supplement), 18 Nov. 1973; Anthony Monahan, "Looking for Monster-San."
Modern Photography, Dec. 1976, 176; "You're Messing about with Our Monsters, Mon!"
MORE, July/Aug. 1976, 57–58; Richard Pollak, "Monster Amok in Newsroom."
Mufon UFO Journal, Oct. 1982, 17; Robert Wanderer, "Critic's Corner—Nessie and UFOs."
Nation, 24 Jan. 1934, 138 : 102–3; "In the Driftway."
National Geographic, June 1977, 758–79; William S. Ellis, "Loch Ness—the Lake and the Legend."
National Geographic World, no. 20 (Apr. 1977): 4–9; "Loch Ness Monster Hunt."
Natural History, 34 (1934): 327–31; William K. Gregory, "Sea Serpents."
———, 34 (1934): 674–76; William K. Gregory, "The Loch Ness 'Monster'."
———, 66 (1957): 183–87, 222; Richard Carrington, "Sea Serpent Riddle of the Deep."
———, 77 (1968): 71–74; Willy Ley, review of Heuvelmans, *In the Wake of the Sea-Serpents* (1968).
———, 89 (1980): 130–31; reprint of William K. Gregory, "Sea Serpents" (1934).
Nature, 16 Dec. 1933, 132 : 921; "The Loch Ness 'Monster'" (from a correspondent).
———, 13 Jan. 1934, 133 : 56.
———, 18 Aug. 1934, 134 : 242; by R.J.
———, 17 Nov. 1934, 134 : 765.
———, 28 Dec. 1968, 220 : 1272; "News and Views—Monsters by Sonar."
———, 11 Jan. 1969, 221 : 201; P. F. Baker, "Monsters by Sonar."

———, 18 July 1970, 227:215–16; "Monsters and Poltergeists" (editorial).
———, 19 Sept. 1970, 227:1277; James M. Piggins, "Monsters and Poltergeists" (letter).
———, 28 Nov. 1970, 228:797.
———, 11 Dec. 1975, 258:466–68; "Naming the Loch Ness Monster."
———, 25 Dec. 1975, 258:655; "Nessiteras Skeptyx."
———, 15 Jan. 1976, 259:75–76; (letters) Gordon B. Corbet and L. B. Halstead et al., with replies by Sir Peter Scott.
———, 9 Dec. 1976, 264:497; Carl Sagan, "If There Are Any, Could There Be Many?"
New Age Journal, Dec. 1984, 47–51; Alice van Buren, "A Postcard from Loch Ness."
New Humanist, May/June 1976, 20–21; Stuart Campbell, "Reasonable-Ness?"
New Scientist, 22 Sept. 1960, 773–75; Maurice Burton, "Loch Ness Monster: A Reappraisal."
———, 27 Oct. 1960, 1144–46; (letters) Denys W. Tucker, Dorothy E. Warren, and Constance Whyte, with reply to Tucker by Maurice Burton.
———, 17 Nov. 1960, 1346–47; (letter) Denys W. Tucker.
———, 24 Nov. 1960, 1413–14; (letters) Peter F. Baker et al. and L. B. Tarlo.
———, 1 Dec. 1960, 1478; (letters) Maurice Burton and Angus Ross.
———, 8 Dec. 1960, 1549–50; (letter) Maurice Burton.
———, 5 Jan. 1961, 51; (letters) Peter F. Baker and Mark Westwood, and Constance Whyte.
———, 19 Dec. 1968, 664–66; Hugh Braithwaite, "Sonar Picks Up Stirrings in Loch Ness."
———, 26 Dec. 1968, 729; Rex Malik, "Is Nessie Now Respectable?" (review of Holiday, *The Great Orm of Loch Ness* [1968]).
———, 9 Jan. 1969; F. C. Odds, "A Load of Rubbish."
———, 23 Jan. 1969; M. Burton, "Verdict on Nessie."
———, 27 Mar. 1969; Maurice Burton, review of Heuvelmans, *In the Wake of the Sea-Serpents* (1968).
———, 25 July 1974, 207–8; Tony Loftas, review of Costello, *In Search of Lake Monsters* (1974).
———, 27 Nov. 1975, 499; "Nessie Fails Again to Surface."
———, 4 Dec. 1975, 585; "Will Rines Patent Nessie?"
———, 18/25 Dec. 1975, 738–39; "The Pictures Show Something, But Is It Nessie?"
———, 18/25 Dec. 1975, 739; "Monster Hoax?"
———, 18/25 Dec. 1975, 739; "What Do Monsters Eat?"
———, 12 February 1976, 346; "New Nessie Sonar Tracks Are Fake."
———, 8 July 1976, 92; Joseph Hanlon, review of Mackal, *The Monsters of Loch Ness* (1976).

———, 21 June 1979, 982; "Dolphin Chase for Nessie."
———, 5 July 1979, 6; "Nessie Hunter Dies."
———, 2 Aug. 1979, 358–59; Dennis Power and Donald Johnson, "A Fresh Look at Nessie."
———, 23 Aug. 1979, 613; Y. Franz Daun, "Nessie" (letter).
———, 17 June 1982, 779; Barry Fox, "Patents—Camera for Dolphins."
———, 24 June 1982, 872; Maurice Burton, "The Loch Ness Saga—A Ring of Bright Water?"
———, 1 July 1982, 41–42; Maurice Burton, "The Loch Ness Saga—A Fast Moving, Agile Beastie."
———, 1 July 1982, 47; Nigel Sitwell, "Loch Ness Dolphins."
———, 8 July 1982, 112–13; Maurice Burton, "The Loch Ness Saga—A Flurry of Foam and Spray."
———, 22 July 1982, 259; Andrew Butler, "Dr. Who?"
———, 5 Aug. 1982, 354–57; Robert P. Craig, "Loch Ness: The Monster Unveiled."
———, 19 Aug. 1982, 516; Sinclair C. Dunnett, Naomi Mitchison, and H. J. Walls, "Loch Ness Monster" (letters).
———, 26 Aug. 1982, 573; Jon Swan, "A Question of Degree."
———, 16 Sept. 1982, 792; Graham Jones, "Likely Tale."
———, 23 Sept. 1982, 861–62; Robert Tullock, "Palm Ness Monster."
———, 17 Feb. 1983, 462–67; Adrian Shine, "The Biology of Loch Ness."
———, 18 Aug. 1983; "Nessie Hunt."
New Statesman, 2 June 1961; Ted Hughes, "Five Ton Phantom" (review of Dinsdale, *Loch Ness Monster* [1961]).
———, 13 July 1984; Alan Brien, "Popular Prejudice" (review of Binns, *The Loch Ness Mystery Solved* [1983]).
New Times, 7, no. 8 (24 Oct. 1976): 76; Lawrence Wright, "Final Tribute—The Monster Shortage."
New York Running News, Dec./Jan. 1984/85; "A Monster Run."
New Yorker, 2 Dec. 1972; review of Baumann, *The Loch Ness Monster* (1972).
The News, no. 2 (Jan. 1974): 13; "Round One to Nessie."
Newsweek, 13 Jan. 1934, 16–17; "Sea Serpent: Shy Monster Evokes Conflicting Tales from Sober Scots."
———, 27 May 1957; "Scotland: The 'Monster' Season."
———, 15 Aug. 1960, 78; "Biology: Where's Nessie?"
———, 4 Aug. 1975, 15; Ann Ray Martin, "Caught in the Act?"
———, 4 Dec. 1978, 26; "Elusive 'Nessie'."
Nineteen (U.K.), no. 13 (July 1984): 48–49; Jane Dowdeswell and Francesca White, "Summer Madness Special!"
Nineteenth Century, Feb. 1934, 220–30; Malcolm Burr, "Sea Serpents and Monsters."
North American Review, Mar. 1934, 237:257–62; P. W. Wilson, "Sea Serpents and Scientists."

Notes and Queries, 17 Mar. 1934, 166:181; "Memorabilia."
———, 3 Sept. 1938, 175:169–70; W. W. Gill, "The Loch Ness Monster."
Now, 4 July 1980, 40–42; Nigel Sitwell, "The Nessie Disciples Cast Another Net."
Observer, 28 May 1961, 30; Denys W. Tucker, "Case for the Monster" (review of Dinsdale, *Loch Ness Monster* [1961]).
———, 14 Dec. 1975, 11; Pearson Phillips, "Nessiteras Absurdum."
Oceanology International, Sept./Oct. 1967, 38–44; Roy P. Mackal, "'Sea Serpents' and the Loch Ness Monster."
Omni, May 1979, 92–95, 123–26; John Chesterman and Michael Marten, "Return to Loch Ness."
———, Aug. 1979, 37; "O.D.—Loch Ness Dolphins."
———, Jan. 1983, 108–13, 120; Karen Ehrlich and Lee Speigel, "Hidden Monsters."
Oryx, Apr. 1962, 6:241–42; D.W.T., review of Burton, *The Elusive Monster* (1961), Dinsdale, *Loch Ness Monster* (1961), Gould, *The Loch Ness Monster and Others* (1934), and Whyte, *More than a Legend* (1957).
———, Oct. 1973, 12:256–57; Richard Fitter, review of Campbell, *The Search for Morag* (1972), and Dinsdale, *Monster Hunt* (1972).
———, Apr. 1975, 13:38–40; Richard Fitter, "Nessie and the Sasquatch" (review of Costello, *In Search of Lake Monsters* [1974], and Witchell, *The Loch Ness Story* [1974]).
Oui, May 1974, 50–52, 82, 112–22; Dan Greenburg, "Japanese Come to Catch Loch Ness Monster."
Outline (Abbott Laboratories, Chicago), 9 May 1969, 1, 4; "Stalk Legendary Sea Monster: Sigma Xi Invitation."
Paris-Match, 5 May or 5 Sept. 1953; Guillaume Hanoteau, "Notre envoyé spécial au Loch Ness a guetté Nessie, le monstre des vacances."
———, 11 Apr. 1959; "Je suis le père du monstre du Loch Ness."
———, July 1976, 51–54; "Maintenant les sonars traquent le monstre."
Passages (Northwest Airlines Magazine), Oct. 1980, 28, 33–35; Jon R. Luoma, "Case Study No. 2—Nessie."
People, 29 Apr. 1959, 51–53; Alex M. Campbell, "Plenty of Loch Ness Monsters!"
———, 17 Aug. 1960, 17–19; Tim Dinsdale, "Scotland's Garbo Has Made a Fine Debut."
———, 10 May 1976, 34–35; Jerene Jones, "Two Loch Ness Observers Say Nessie Is Real, But Please Don't Call It a Monster."
Photographic Journal, Mar. 1970, 90–97; Tim Dinsdale, "The Potential of Photography at Loch Ness."
———, Apr. 1973, 162–65; Tim Dinsdale, "The Rines/Edgerton Picture."
———, Jan./Feb. 1976, 1–4; Tim Dinsdale, "Loch Ness '75."

Photomethods, Sept. 1977, 10, 14; William G. Hyzer, "Taking the Odds on Nessie."
———, Oct. 1977, 17, 68; "Zeroing in on Nessie."
Le Point (Paris), 1 Dec. 1975, 119; Marie-Claude Descamp, "Le monstre prend le frais."
———, 30 July 1979, 38–39; "Deux dauphins pour déloger Nessie."
Popular Mechanics, Sept. 1934, 62:398–401, 118A; "Is There a Sea Serpent?"
———, Sept. 1976, 146:92–95, 144, 146, 148; Dan Cohen, "Seven Great Quests of Man."
Popular Photography, 85, no. 1 (1979): 70; "Finny Photographers."
Popular Science, Nov. 1966, 112–15, 212, 214, 216; David Scott, "New Evidence Spurs Hunt for Loch Ness Monster."
Popular Science Monthly, 176, no. 6 (1961): 69–71; Gardner Soule, "From the Loch Ness Monster to the Giant Squid."
Ports of Virginia News Letter–Sailing Schedule (Hampton Roads Ports), July 1969, 4; "Containerized Sub Will Hunt Monster."
Probe, July 1980, 80–85; Ben Singer, "Loch Ness: Unsolved, But Not Unfounded."
Probe the Unknown, May 1976, 33–39, 58–60; Catherine Coggins, "Loch Ness Monster: Fact or Fable?"
Proceedings of the Challenger Society, 4 (1970): 91–92; D. G. Tucker and D. J. Creasey, "Some Sonar Observations in Loch Ness."
Proceedings of the Linnean Society of London, 8 Nov. 1934, part 1, 7–12.
Publishers Weekly, 21 Feb. 1972; review of Baumann, *The Loch Ness Monster* (1972).
———, 12 Jan. 1976, 42; Paul S. Nathan, "Rights and Permissions—Strange Creatures" (preview of Mackal, *The Monsters of Loch Ness* [1976]).
Punch, 1 Aug. 1962; "Richard Fitter Considers the Loch Ness Monster."
———, 5 May 1971, 604; "Loch Ness: The Law Steps In."
———, 13 Oct. 1976; Derek Cooper, "The Monstrous Regiment" (review of Dinsdale, *The Leviathans* [1976], and Mackal, *The Monsters of Loch Ness* [1976]).
Purple Martin News (Griggsville, Ill.), 27 Dec. 1977; John V. Dennis, "The Critical List: Unloching the Monster's Secret."
———, 24 Apr. 1978; John V. Dennis, "The Critical List: He Has Eyewitness Sketch of the Loch Ness Monster: Is This the Loch Ness Monster?"
Pursuit, 2, no. 1 (1969): 12; "Sea-Cows and Water-Horses."
———, 2, no. 4 (1969): 72–73; "That Damned Bone."
———, 4, no. 2 (1971): 42–43; Jack A. Ulrich, "'Nessie' Is Alive and Well and Living in Urquhart Bay."

———, 4, no. 4 (1971): 95; "More on Jack Ulrich's Loch Ness Photographs."

———, 7, no. 2 (1974): 46–48; Tim Dinsdale, "Loch Ness 1972: The Rines/Edgerton Picture."

———, 7, no. 3 (1974): 68–70; R. S. Wheldon and S. A. Kerr, "The Population Density of Monsters in Loch Ness."

———, 8, no. 3 (1975): 67–68; Marty Wolf, "An Interview with Tim Dinsdale."

———, 9, no. 3 (1976): 56–60; R. H. Rines, C. W. Wyckoff, H. E. Edgerton, M. Klein, and J. M. Breece III, "Fresh Water Monsters."

———, 11, no. 1 (1978): 2–6; Joel A. Strasser, "Loch Ness Update, 1977."

———, 11, no. 3 (1978): 126; David Weidl (letter), and editorial response.

———, 11, no. 4 (1978): 151–52; "Postscript: Phantom Sea Monsters, Too?"

———, 11, no. 4 (1978): 153–58; Robert J. M. Rickard, "The Shiels Nessie Photographs."

———, 15, no. 2 (1982): 50, 51, 56, 63; Sydney Wignall, "Morag of Morar."

———, 15, no. 2 (1982): 78–79; Sabina W. Sanderson, review of Dinsdale, *Loch Ness Monster* (1982).

———, 16, no. 2 (1983): 78, 88; Joseph W. Zarzynski, "Loch Ness 'Monster's' Fiftieth Birthday is Quietly Unobserved."

Quick, 5 Feb. 1976, 1–7; Harvey T. Rowe, "Was an Nessie wahr ist—und was nur witzig war."

Radio Times (U.K.), 244, no. 3177 (29 Sept.–5 Oct. 1984); "Together Ness."

Ranger Rick (National Wildlife Federation), Mar. 1985, 15; "Is the Loch Ness Monster Just a Lot of Hot Air?"

Reader's Digest, Nov. 1957; James Thurber, "*Is* There a Loch Ness Monster?" (condensed from *Holiday*, Sept. 1957).

———, Feb. 1967, 86–90; David Scott, "Closing in on the Loch Ness Monster" (from *Popular Science*, Nov. 1966).

———, Feb. 1977, 120–24; James Stewart-Gordon, "In Pursuit of the Loch Ness Monster."

———, May 1983, 167–72; Virginia Morell, "He Hunts for Living Dinosaurs" (condensed from *Equinox*, Sept. 1982).

La Recherche, 7, no. 65 (Mar. 1976): 278–81; Roger Tippett, "Loch Ness: Un monstre mal nourri."

Review of Reviews (London), Jan. 1934, 85 : 20–22; "A Monster in Loch Ness."

Saga, Jan. 1967, 33 : 20–23, 78–79, 81–82; Ivan T. Sanderson, "'Monster' Hunting."

———, Oct. 1970, 22–25, 92, 94; Elwood D. Baumann, "Is 1970 the Year We Unlock the Mystery of the Loch Monsters?"

———, Sept. 1976, 20–23, 56–62; James Natal, "Loch Ness Monsters. . . ."

The Saturday Book, 13 (1953): 104–14; Anna McMullen, "A Census of Sea-Serpents."

Saturday Evening Post, 8 Mar. 1947, 219:22ff.; Ivan T. Sanderson, "Don't Scoff at Sea Monsters."

Saucer News, Sept. 1964, 11:11; "Loch Ness Monster Seen Again and Caught on 'Sonar'."

———, June 1966, 13:33; "Film of Loch Ness Monster Judged to Be Genuine."

———, Summer 1967, 14:10–11; Daniel Cohen, "The Loch Ness Monster Lives!"

School Library Journal, 15 Feb. 1973; Michael Cart, review of Baumann, *The Loch Ness Monster* (1972).

———, Feb. 1980; Linda Blaha, review of Rabinovich, *The Loch Ness Monster* (1979).

Science, 15 Nov. 1968, 162:787–88; Joel W. Hedgpeth, "Elusive Specimens" (review of Heuvelmans, *In the Wake of the Sea-Serpents* [1968]).

———, 9 Jan. 1976, 191:54; "'Nessie': What's in an Anagram?"

———, 19 May 1978, 200:722–23; J. Richard Greenwell, "An Endothermic 'Nessie?'"

———, 13 Apr. 1979, 204:159; "Dolphins to Look for Nessie."

———, 13 July 1979, 205:183–85; W. H. Lehn, "Atmospheric Refraction and Lake Monsters."

Science (France), 17 July 1976, 51–54; "Maintenant les sonars traquent le monstre."

Science Books, May 1974; review of Baumann, *The Loch Ness Monster* (1972), and Campbell, *The Search for Morag* (1972).

Science Digest, Oct. 1961, 61–63; "Loch Ness Monster Again."

———, July 1962, 68–73; "New Light on the Loch Ness Monster" (from *Illustrated London News,* 24 Feb. 1962).

———, Jan. 1967, 20–22; Daniel Cohen, "Looking for the 'Monster'."

———, Feb. 1971, 8; "'Monsters' in Another Loch."

———, Sept. 1971; "How to Catch a Monster."

———, Jan. 1972, 60–64; Mark Roberts, "British Science Vacation."

———, Oct. 1972, 6–7; John T. LaFleur, "Hunting the Loch Ness Monster" (letter).

———, May 1974; "Nessie Is Back and Looking Fit."

———, Mar. 1976, 8–9, and inside front cover; "Loch Ness Monster."

———, Aug. 1976, 91–92; review of Mackal, *The Monsters of Loch Ness* (1976).

———, Feb. 1978, 66–75; Dennis L. Meredith, "Search at Loch Ness."

Science Journal, Dec. 1970, 6:21–22; "Infrared Camera to Track Down 'Nessie'."

Science News, 20/27 Dec. 1975, 108:391; "Nessie: New Name, Same Old Controversy."
———, 17 Apr. 1976, 109:247–48; "The Case for the Loch Ness Monster."
———, 5/12 June 1976, 109:359; "Loch Ness Search Sponsored by *Times.*"
———, 14 Aug. 1976, 110:103; "Ancient Stoneworks Found in Loch Ness."
———, 31 Mar. 1979, 115:200; "Smile for the Dolphins, Nessie!"
———, 18 Aug. 1979, 116:122–23; Marcia F. Bartusiak, "Will the Real Nessie Please Stand Up?"
Sciences et Avenir, no. 435 (1983); "Actualités—Loch Ness: Le monstre releve la tête."
Scientific American, Feb. 1935, 152:67; "Our Point of View—Too Much Scotch?"
———, 219, no. 6 (1968): 129; review of Heuvelmans, *In the Wake of the Sea-Serpents* (1968).
Scotland (S.M.T.) Magazine, Sept. 1947, 42–44; J. W. Herries, "The Loch Ness Mystery."
———, June 1957, 42; Robert Wotherspoon, "Nessie" (review of Whyte, *More than a Legend* [1957]).
———, July 1957, 40–42; Isobel Knight, "The Annals of the Loch Ness Monster."
———, Apr. 1972, 25; Edyth Harper, "Nessie—Fact or Fiction?"
Scots Magazine, Sept. 1948, 454–58; D. D. C. Pochin Mould, "The Beast in the Loch."
———, May 1962, 95–100; A. M. Campbell, "No, Dr. Burton!"
———, Oct. 1983, 55–63; Roy Fraser, "The Nessie Mystery Solved?"
———, Jan. 1984, 440–41; A. M. Watt, "Monster Tales" (letter).
———, Apr. 1984, 98; Margorie Wilson, "Monster Sighting" (letter).
———, Apr. 1984, 100; George Mackenzie, "Abbot's Tale" (letter).
Scottish Field, Feb. 1960, 48–49; Maurice Burton, "What Is the Loch Ness Monster?"
———, Nov. 1960, 33; Maurice Burton, "New Line on the Monster."
———, Apr. 1975, 44–45; A. Ruari Grant, "The Loch Ness Monster—Is It a Ghost?"
———, June 1975; E. M. Patterson, and Stuart Campbell, "Monster Theory Rejected" (letters).
———, Dec. 1975, 29; Rosemary Hamilton, "Loch Ness Monster."
Sea Secrets, 1966 (later than June); "Sea Winds—More about Loch Ness Monster."
Seahorse (Hydro Products, Dillingham Corporation), 4, no. 1 (Feb. 1970): 1, 2; "'Nessie' Eludes Pursuers in Latest Monster Hunt."
Senior Scholastic, 18 May 1976, 28–29; "World History—Loch Ness Monster: Fact or Fiction?"

Skeptical Inquirer, 7, no. 2 (Winter 1982–83): 42–46; Steuart Campbell, "The 'Monster' Tree-Trunk of Loch Ness."

———, 7, no. 4 (Summer 1983): 72–73; James F. Waters, Stuart Lucas, and J. Erik Beckjord, "Loch Ness photos" (letters)

———, 9, no. 1 (Fall 1984): 91–92; Henry H. Bauer, "About Loch Ness."

———, 9, no. 1 (Fall 1984): 92; Steuart Campbell, "Null Hypothesis on Nessie?"

Skin Diver, Oct. 1972, 56, 86; Arthur Bourne, "Eight Hundred-Foot Descent into the Monster's Lair."

———, 28, no. 9 (Sept. 1979): 70–71; Robert Davidsson, "Dolphins for Loch Ness."

Smithsonian, June 1976, 96–105; John P. Wiley, Jr., "Cameras, Sonar Close in on Denizen of Loch Ness."

———, Aug. 1976; (letters) Cindy Butler Jones, Virginia Vilas et al., and K. P. Carroll.

Le Soir Illustré (Brussels), 14 Apr. 1966, 20–25; Tim Dinsdale, "Voici les photos véridiques du monstre du Loch Ness."

Spectator, 20 July 1934, 94; W. T. Calman, "The Monster's Credentials" (review of Gould, *The Loch Ness Monster and Others* [1934]).

———, 15 Dec. 1950, 686; Sir Harold Spencer Jones, "The Flying Saucer Myth."

———, 5 Apr. 1957, 439–40; Strix, "Monster in Aspic" (review of Whyte, *More than a Legend* [1957]).

———, 28 Feb. 1969, 274; Norman Collins, "The Kraken Lives" (review of Heuvelmans, *In the Wake of the Sea-Serpents* [1968]).

———, 6 Dec. 1975; Richard Luckett, "Loch Ness—The Aging Hypothetical."

Spiegel, 31 May 1976, 226–29; "Flossen aus grauer Vorzeit?"

Spiegel (Hamburg), 30 July 1979; "Seeungeheuer—optische Tauschung."

Sports Illustrated, 51, no. 6 (6 Aug. 1979): 9–10; Jerry Kirshenbaum, "Scorecard: Busy Molly."

Spotlight (Field Enterprises Educational Corporation), 50, no. 1 (11 Jan. 1969): 2–3; "Sonar Findings May Be Evidence of 'Monster' in Loch Ness."

———, 50, no. 14 (12 July 1969): 1, 8; "PR, Viperfish Plumb for Nessie Antics."

Stern, 15 Sept. 1977; Hans Heinrich Ziemann, "Wie das Monster seine Leute ernahrt."

Technology Review, Mar./Apr. 1976, 2; "Jumping into the Loch."

———, Mar./Apr. 1976, 10–12; Dennis Meredith, "The Loch Ness Press Mess."

———, Mar./Apr. 1976, 25–40; Robert H. Rines, Charles W. Wyckoff, Harold E. Edgerton, and Martin Klein, "Search for the Loch Ness Monster."

———, Oct./Nov. 1976, 19–20; D.M., "Loch Ness '76: Sonar and Surprises."

———, Dec. 1976, 44–57; Martin Klein and Charles Finkelstein, "Sonar Serendipity in Loch Ness."

———, June 1977, 70–71; D.M., "Another Go at Loch Ness."

———, Oct./Nov. 1977, 73; Sara Jane Nustadtl, "Monster Mash" (review of Meredith, *The Search at Loch Ness* [1977]).

———, June/July 1979, 14–16; Robert H. Rines and Howard S. Curtis, "Loch Ness: The Big One Got Away—Again."

Télé Magazine (Paris), 28 Mar. 1962, 19; Monique Lefebvre, "Nessie, monstre du Loch Ness. . . ."

La Terre et la Vie (Paris), Mar. 1934; "Le 'monstre' du Loch Ness."

Das Tier (Frankfurt am Main), Apr. 1963, 10; S. A. Barnett, "Das Loch-Ness Problem."

Time, 3 May 1937, 20; "Again, Nessie."

———, 29 June 1942, 31–32; "All's Well that Ends Well."

———, 20 Nov. 1950, 30; "Monster Rally."

———, 8 Oct. 1951; "Monster on Trial."

———, 27 Dec. 1968, 30; "Marine Biology—Clue to the Loch Ness Monster."

———, 20 Nov. 1972; "Myth or Monster?"

———, 12 Jan. 1976, 39–40; "Nessie's Return."

———, 21 June 1976, 76; "Coverage in Depth."

———, 12 July 1976, 2; Richard Swerdlow and Genevieve Robbins, "Him? Her? It?" (letters).

Times Literary Supplement, 4 Aug. 1966; "Water Monsters" (review of Dinsdale, *The Leviathans* [1966]).

———, 30 Jan. 1969, 112; "A Giant Worm?" (review of Holiday, *The Great Orm of Loch Ness* [1968]).

———, 10 Apr. 1969, 392; "Crypto-zoology" (review of Heuvelmans, *In the Wake of the Sea-Serpents* [1968]).

———, 27 July 1973; "Laidly Worms" (review of Holiday, *The Dragon and the Disc* [1973]).

Trail Prints (Adventures Club, Chicago), Mar.–Apr. 1969, 1; "Quotable Quote—Roy Mackal."

———, July–Aug. 1970; "Nessie, Where Are You?"

Travel (U.S.), June 1970, 60–62; Barbara Bell, "Scotland's Monster Lake."

TV Guide, 11 Dec. 1976, 33–36; Alan Coren, "A Very Tragic Story, Including Five Big Dance Numbers."

TV Times, 31 July 1969, 3–6; "Operation Nessie," including Gavin Maxwell, "I Saw the Secret of the Loch."

The Unexplained (U.K.), no. 10 (1980): 190–93; "Loch Ness Photo File."

———, no. 11 (1980): 214–17; Adrian Shine, "Rumours, Legends and Glimpses."

———, no. 12 (1980): 226–29; Adrian Shine, "Sounding Out the Sightings."

———, no. 13 (1980): 241–45; Adrian Shine, "A Very Strange Fish?"

———, no. 14 (1981): 264–68; Adrian Shine, "To Catch a Monster."

University of Chicago Magazine, Sept./Oct. 1968, 9–17; C.K., "The Quest at Loch Ness."

US, 24 July 1979, 6; "Dolphins Will Search for Evasive 'Nessie'."

Variety, 4 July 1973, 2; "Loch Ness Monster in Japanese Revival."

La Vie des Bêtes (Paris), no. 229 (Aug. 1977): 8–11; François Gohier, "L'énigme du Loch Ness; un entretien avec Peter Scott."

Vokrug Sveta (Moscow), Jan. 1958, 50–52; I. Akimushkin.

Weekend (Sydney), 6, no. 20 (2 Jan. 1960): 28–29; "He Fought the Horror of Loch Ness."

———, 7, no. 5 (17 Sept. 1960): 38; "Monsters Galore."

———, 8 July 1981; "Has Nessie Got Worms?"

Welt, 4 Sept. 1976; Willy Lutzenkirchen, "Schottische Monster in Torfsee."

Wildlife, Mar. 1976, 102–9; Nigel Sitwell, "The Loch Ness Monster Evidence."

———, Mar. 1976, 110–11; Sir Peter Scott, "Why I Believe in the Loch Ness Monster."

Woman's Weekly (London), 1977, 6–7, 22–24; "Out of Town, and Nessie."

World Fishing, Jan. 1955, 30; "Animal, Vegetable. . . ?"

World Press Review, 29, no. 10 (Oct. 1982): 55; Robert P. Craig, "Explaining 'Nessie'."

Yankee, June 1971, 74–79, 169; Richard W. O'Donnell, "Now We Know What the Loch Ness Monster *Sounds* Like!"

———, May 1972, 36: 204–5; Richard W. O'Donnell, "What Do *You* Think?"

Zeit, 31 Dec. 1976; Manfred Sack, "Nessie, wo bist du?"

Zetetic, 2, no. 1 (Fall/Winter 1977): 110–21; Bernard Heuvelmans, review of Mackal, *The Monsters of Loch Ness* (1976).

Zetetic Scholar, no. 6 (July 1980): 17–29; J. Richard Greenwell and James E. King, "Scientists and Anomalous Phenomena: Preliminary Results of a Survey."

———, no. 7 (Dec. 1980): 30–42; Henry H. Bauer, "The Loch Ness Monster: A Guide to the Literature."

———, no. 10 (Dec. 1982): 160–61; Henry H. Bauer, review of Dinsdale, *Loch Ness Monster* (1982).

———, forthcoming; Henry H. Bauer, review of Binns, *The Loch Ness Mystery Solved* (1983).

NEWSPAPERS

(section, page, and column, where known, are given in parentheses after the date)

Glasgow Herald(1933–69)

1933: Oct. 12, 27; Nov. 13(7a), 25(11d); Dec. 5(9d), 6(13d), 8(13e), 11(7a), 12(2g), 13(11a, 11b), 15(11e), 16(10g), 18(11e), 19(9e), 20(10e), 27(9b), 28(9c), 30(2g)
1934: Jan. 1(4d, 9c), 2(7c), 5(7e), 8(11f), 10(11d), 11(9c), 12(12b), 13(9f), 15(13f), 16(3f), 17(12c), 18(11c), 19(12f); Feb. 10(9c), 13(2c), 27(6g); Mar. 3(12c), 5(13b), 6(7g), 29(10d); Apr. 2(4f), 3(6e), 4(10c), 17(3f), 20(14c), 26(11f); May 5(12c), 8(6f), 10(8c), 15(10e), 16(12f), 21(8a), 23(11b), 28(19f), 31(12e); June 2(11d), 15(9a), 18(11f), 25(11g); July 3(10f), 6(9b), 7(9b), 11(9e), 14(9a), 16(7f), 17(7g), 20(9e), 24(7d), 25(9g), 30(9b); Aug. 3(9b), 6(7d), 7(7c), 9(9f), 14(7d), 15(11e), 28(7a); Sept. 1(7g), 5(14g), 15(8a), 17(9g), 25(7b); Oct. 2(6e), 4(10f), 5(10g), 15(9g), 27; Nov. 17(9g); Dec. 8(7e), 29(5f)
1935: Jan. 15(10c), 21(9f); Feb. 8(11e); Apr. 22(11b), 26(13g); May 19(11d), 24(8g), 25(9g); June 12(11d); July 13(9b), 22(9g); Aug. 6(5e), 19(9b), 22(9b); Sept. 9(9a); Dec. 24(7a)
1936: Apr. 30(9d); May 11(13d); June 22(11c); Sept. 12(9g), 22(4g), 30(9e); Oct. 1(9e); Nov. 2(7f), 24(9e)
1937: Mar. 16(9c); Apr. 1(6a); July 27(7a); Nov. 25(9b)
1938: Mar. 3(9c), 7(11b)
1939: May 26(11c); Aug. 14(9b), 16(9a), 30(9a)
1944: Apr. 5(5c)
1945: May 22(2a)
1947: May 2(5d), 3(L), 15(5g), 16(4g)
1948: Jan. 21(2f); July 16(7g); Dec. 24(5f)
1950: Apr. 29(7b); June 21(6d); Aug. 30(7d)
1951: Nov. 17(6f)
1952: Feb. 29(5e); Apr. 29(5c); Aug. 22(7a)
1953: Dec. 16(7c)
1954: July 12(9c); Aug. 10(4e), 14 (6b), 17(6c); Dec. 4(9c)
1955: June 27(7f)
1957: Mar. 12(7d); 19(6c); May 8(10f); June 17(7g); Dec. 30(6f)
1958: Apr. 15(3g); May 15(12f), 21(10c); June 16(8g)
1960: June 16(9b); July 4(8g); Nov. 28(8d)
1961: June 8(12e); July 29; Aug. 14(5c)
1962: Oct. 15(10d)
1963: Feb. 25(8h); Apr. 22(6h); June 3(3g), 10(9e), 14(6d); July 10(5d); Aug. 17(1d)
1964: Mar. 21(1d), 30(9d); May 15(13g)
1965: May 17(6b); July 8(1e); Aug. 11(5b); Sept. 15(11g); Dec. 29(5a)

1966: Jan. 22(13f); June 14(1e); July 15(7a, 8a); Sept. 20(7), 24(7h), 27(7c)
1967: Jan. 3(7e); Feb. 16(5b); Mar. 13(1b); Apr. 10(1e); May 15(10c); July 21(1e); Aug. 24(16c); Sept. 29(8d)
1968: Apr. 20(13d); July 8(1d); Dec. 19(1a), 20(18d)
1969: Oct. 20(18)

Inverness Courier (1930–39)

1930: Aug. 29(letter)
1933: May 2, 12(5:v), 23(4:v), 30(4:v); June 2(5:ii), 9(5:v); August 4(5:i), (5:i), 8(5:iv), 11(5:i, ii), 15(6:ii), 18(5:i–ii), 25; Sept. 5, 12, 15(4:vi), 26(4:iii; 5:iii); Oct. 3(5:iv–v), 10(?:v–vi), 13(3:v), 20(5:vi; 6:i), 24(4:iv; 5:ii–iii; 5:iv), 27(6:i), 31(5:i); Nov. 3(5:iii; 7:ii), 14, 17(5:ii), 21(5:iv); Dec. 5(6:vi), 8(5:ii), 12(5:i; 5:vi; 6:i), 15(5:vi), 19(5:ii–iii; 5:v), 22(5:ii–iii), 26(5:i–ii), 29(4:v–vi; 5:i; 5:v)
1934: Jan. 2(5:i; 5:vi; 6:iv), 5(4:v–vi; 5:iii–iv; 6:i–ii), 9(4:ii–iii; 4:v–vi; 3:i; 5:iii–iv), 12(3i; 5:iii–iv), 16(4:v–vi; 5:ii), 19(5:ii), 23(5:vi), 26(4:vi; 5:iv), 30(5:ii; 5:iii); Feb. 2(4:vi), 13(3:vi; 5:ii), 16(5:v), 20(5:vi), 27(5:iii); Mar. 2, 6(4:iv), 30(4:vi; 5:iv); Apr. 3, 6(poem by A.D.L.), 20; May 4, 8, 22; June 1, 5, 26; July 3, 6, 10, 13, 17, 24, 27, 31; Aug. 3, 7, 14, 24, 28; Sept. 11, 28; Oct. 2, 5, 8, 26
1935: Jan. 15, 22; Feb. 8; Apr. 23; May 24; June 21; Aug. 6, 20, 27, 30; Sept. 3, 10, 13
1936: Mar. 24, 31; June 23; Sept. 22, 29; Oct. 17; Nov. 24
1937: Feb. 12; Mar. 16; Nov. 9, 26
1938: July 1, 12; Aug. 19, 23, 26, 30; Sept. 2
1939: Apr. 7; May 20; June 23; July 4; Aug. 8, 15, 18

New York Times (1933–79)

1933: Dec. 9(6:1), 22(23:8), 31(1:2)
1934: Jan. 1(25:6), 2(10:8), 3(14:8), 4(16:3), 5(20:4), 6(10:2), 7(5:4), 9(20:3), 13(12:4), 14(2:4), 17(2:2), 19(21:3), 23(18:3), 27(12:6), 30(21:6), 31(16:4); Feb. 11(3:3); Mar. 2(21:3), 4(2:3), 14(23:8); Apr. 5(12:6), 21(9:2), 22(II, 1:7), 24(26:5), 25(17:5, 20:4), 26(46:5); May 8(6:3); June 4(19:4); Sept. 3(12:4), 5(1), 25(23:1), 29(17:6); Oct. 17(9:5); Nov. 11(VIII, 5:2)
1935: May 26(10:4)
1937: June 27(21:5)
1948: Apr. 3(17:7)
1950: Nov. 10(3:7)
1952: Jan. 13(IV, 10:6)
1957: May 12(17:1)
1960: July 24(12:1), 30(16:5)
1962: Mar. 27(24:6)

1966: Feb. 21(3); Sept. 21(17 : 1), 26(11 : 1)
1967: May 14(66 : 4)
1968: May 29; Nov. 10(VII, 46–48)
1969: Jan. 27(16 : 1); Dec. 14(34)
1970: Sept. 24(31 : 1)
1971: Apr. 25(72 : 1)
1972: Jan. 16(X, 3); Apr. 2(15 : 1); Nov. 3(7 : 3)
1973: Aug. 18(22 : 3)
1974: Feb. 3(X:1, 11), 17(X, 4), 24(XII, 11); June 2(X, 31 : 1)
1975: Nov. 23(4 : 1); Dec. 2(8 : 6), 5(1 : 3), 11(91 : 4), 14(10 : 3), 19(78 : 5)
1976: Apr. 8(1 : 3), 11(IV, 7 : 1), 15(32 : 5), 21(36 : 4); May 28(1 : 5), 30(IV, 1 : 3); June 4(I, 3 : 3), 5(8 : 4), 6(1 : 5, 10), 7(4 : 4), 10(2 : 4, 36 : 3), 11(11), 12(2 : 3), 13(IV, E, 8 : 6), 14(18 : 2), 16(40 : 1), 17(34 : 5), 19(21 : 1), 20(IV, 7 : 4), 25(26 : 3), 27(10 : 3), 30(36 : 4); July 13(10 : 1), 18(IV, 7 : 1); Aug. 1(VI, 55), 3(8 : 3), 8(IV, 5 : 2); Oct. 10(15 : 1); Nov. 5(II, 6 : 6); Dec. 6(2 : 3), 12(IV, 6 : 5)
1977: June 12(18 : 1); Oct. 9(49 : 3)
1979: Mar. 22(2 : 3); June 28(16 : 6); Aug. 14(III, 1 : 1)

Times (London) (1933–83)

1933: Dec. 8(14g), 9(13–14), 11(15, 16), 12(14), 13(10), 15(14f), 18(9), 20(8), 22(6), 23(6), 28(6), 29(6), 30(6)
1934: Jan. 1(9, 12), 2(8, 14), 3(6), 4(10), 5(8), 6(6, 10e), 18(13), 22(8); Apr. 20(6); May 2(10); June 29(10); July 16(13); Aug. 9(10, 14), 11(6), 22(6); Sept. 25(8), 28(11); Oct. 5(12)
1938: May 9(15), 16(15), 20(17), 23(15), 27(17), 28(15), 31(12); June 1(10), 3(10), 6(6), 7(8), 11(13), 14(11), 17(12), 18(15); Aug. 29(10); Sept. 21(6)
1950: Aug. 30(3b)
1954: Dec. 4(6b)
1958: June 16(8f)
1960: June 30(7c); July 6(8d), 7(13g); Aug. 14(17)
1961: Apr. 21(15f), 25(15e), 28(15g), 29(9f)
1962: May 30(14f)
1963: May 30(5e); June 14(12e)
1966: Dec. 22(11a)
1967: July 22(2f)
1968: Oct. 2(9a); Dec. 21(6b), 28(7f)
1969: May 20(6h), 26(2h); July 14(8f), 17(8a); Aug. 29(9f)
1970: July 9; Sept. 23, 28(9e); Oct. 8(11f), 21(11f); Nov. 25; Dec. 5(13g)
1971: Apr. 26(13d)
1972: Apr. 1(1b), 2, 3(2h); May 12(14h)
1973: Aug. 27(2e)
1975: July 8(2c); Nov. 24(2b), 25(3d), 30; Dec. 2(1h), 3(16d), 4(3e),

5(4g), 6(12h), 11(4h), 12(17g), 13(15e), 15(11g), 18(13g), 19(15g), 27(8c), 31(11f)
1976: Jan. 2(11d), 22(6d), 28(4d); Feb. 29(19); Mar. 14(12g), 15(9a); May 12(16g); Aug. 2
1978: Aug. 11(2h), 12(2g)
1979: Mar. 23(19:6); Apr. 4(1), 5(8); Aug. 1(3:3)
1980: Feb. 16(12a); Mar. 6(17g)
1982: Jan. 2(3:2)
1983: Aug. 9, 10(cartoon), 11, 15, 23(letter), 27

MISCELLANEOUS

Academy of Applied Science Monograph No. 1. Belmont, Mass.: Academy of Applied Science, 1972, 104 pp.
———, *Report to the Membership,* ed. Howard S. Curtis, Belmont, Mass.: Academy of Applied Science, 1982, 40 pp.
Campbell, Elizabeth Montgomery. *Report: Loch Morar Survey 1970.* 39pp.
Carruth, J. A. *Loch Ness and Its Monsters.* Fort Augustus: Abbey Press, 1938, 26pp.; 5th ed., 1960; 6th ed., 1963, 24pp.; 8th ed., 1970, 24pp.; 9th ed., 1971, 24 pp.
Dinsdale, Tim. *Operations Newsletter,* Loch Ness Association of Explorers, 1972–73, 1973–74, summer 1976.
———. *Commentary,* Loch Ness Association of Explorers, 1(5 June 1978), 2(10 Aug. 1978), 3(1 June 1979), 4(14 Aug. 1979), 5(1980), 6(29 Apr. 1982), 7(2 May 1983).
———. *Progress Report, the Smith Film,* Loch Ness Association of Explorers, 19 Sept. 1977.
Grimshaw, Roger, and Paul Lester. *The Meaning of the Loch Ness Monster.* University of Birmingham, Centre for Contemporary Cultural Studies, 1976, 42pp.
Harmsworth, A. G. *The Loch Ness Monster Exhibition,* Drumnadrochit Hotel, 1980, 26pp.
———. *The Mysterious Monsters of Loch Ness.* St. Ives (Huntingdon, Cambs.): Colourmaster International, Photo Precision Ltd., (ca. 1980), 32pp.
Hoy Organization (Box 57, Paducah, Ky.). Press release, 14 Jan. 1977.
Hugli, Paul. *Big Foot, Lochness and other Mysterious Monsters.* Cincinnati: Pamphlet Publications, 1978, 52pp.
James, David. *Loch Ness Investigation.* London: Loch Ness Phenomena Investigation Bureau, (ca. 1968), 40pp.
———. *Annual Report, Loch Ness Investigation.* 1966, 1967, 1968, 1969.
———. "Introduction," *Report on a Film Taken by Tim Dinsdale,* (ca. 1966).
Joint Air Reconnaissance Intelligence Centre. *Photographic Interpreta-*

tion Report: Loch Ness. London: H. M. Stationery Office, 1966.

Lane, W. H. *The Home of the Loch Ness Monster*. Edinburgh: Grant and Murray, 1934, 18pp.

LeBlond, Paul H., and John Sibert. *Observations of Large Unidentified Marine Animals in British Columbia and Adjacent Waters*. Institute of Oceanography, University of British Columbia, manuscript rept. no. 28, June 1973.

Loch Morar Expedition. *'76 Report*. Egham (Surrey): Morar Expedition, 1976, 8pp.

Loch Morar Survey. *1972 Report*.

Loch Ness and the Monster. Newport (Isle of Wight, U.K.): J. Arthur Dixon, 1971, 50pp.

Loch Ness Country by Car. Norwich: Jarrold and Sons, (ca. 1977), 34pp.

Loch Ness and Morar Project. *Report—1980*, 12pp.

———. *Loch Ness Project Report*, 1983.

MacRae, Jim M. *Handbook—Loch Ness Monster*. Inverness, 1979.

The Mysterious "Monster" of Loch Ness. Fort Augustus: Abbey Press, 1934, 20pp.

Nessie—My Own Story. Newtongrange (Midlothian): Lang Syne Publishers, 1979, 40pp.

Nessletter. Ness Information Service (R. R. Hepple, Huntshieldford, St. Johns Chapel, Bishop Auckland, Co. Durham, England DL13 1RQ): 6 issues p.a., no. 1 (Jan. 1974).

Oudemans, A. C. *The Loch Ness Animal*. Leyden: E. J. Brill, 1934, 14pp.

Owen, William. *The Great Glen and Loch Ness*. Norwich: Jarrold and Sons, 1973, 36pp.

———. *Loch Ness Revealing Its Monsters*. Norwich: Jarrold and Sons, 1976, 33pp.

———. *Scotland's Loch Ness Monster*. Norwich: Jarrold and Sons, 1980, 36pp.

Robertson, Barrie. *Loch Ness and the Great Glen*, (ca. 1972), 28pp.

Searle, Frank. *Quarterly Newsletter*, Loch Ness Investigation, Apr. 1977–Dec. 1983.

———. *Loch Ness Investigation—What Really Happened*, (ca. Oct. 1983), 38pp.

———. *The Story of Loch Ness: A Handbook for Nessie Hunters*, 1977, 16pp.

Secrets of Loch Ness, no. 1. New York: Histrionic Publishing Company, 1977, 76pp. Described as an annual publication, but no further issues have appeared. In 1979 an inquiry sent to the cited mailing address was returned as undeliverable.

Time Off (Princeton, N.J., newspaper supplement), week of 28 Feb. 1984, pp. 1, 3, 18; Pam Hersh, "Filmmaker Seeking Key to Loch Ness Mystery."

Welcome to the Highlands (publ. Inverness), 1984, 7; "Nessie—Is She Fact or Fiction?"

Whyte, Constance. *The Loch Ness Monster.* Inverness: Melven Brothers, 1951.

Whyte, Jean M. *The Loch Ness Monster and Its Background: A Select Bibliography.* London: School of Librarianship, Ealing Technical College, 1973, 8pp.

Witchell, Nicholas. *Loch Ness and the Monster.* Newport (Isle of Wight, U.K.): J. Arthur Dixon, 1975, 32pp.; rev. ed. 1976.

Index

(LNM = Loch Ness monster)

Note on the Author

Henry H. Bauer is professor of chemistry and science studies at Virginia Polytechnic Institute and State University, where earlier he served for eight years as dean of the College of Arts and Sciences. A native of Vienna, Austria, Bauer has also held positions at the universities of Sydney (Australia), Michigan, Southampton (England), and Kentucky. His primary area of research has been chemistry, on which he has published several books and nearly a hundred articles. More recently, however, he has focused most of his attention on the interaction between science and culture, and especially on those controversies—like the ones surrounding Loch Ness—that deal with pseudoscience and trans-scientific matters. His latest book is *Beyond Velikovsky: The History of a Public Controversy* (University of Illinois Press, 1984).